THE MAGPIE'S GUIDE TO MONTALCINO

LAURA R. GRAY

Constellations Press

First published in 2024
by Constellations Press

Typeset by Constellations Press

All rights reserved
© Laura R. Gray 2024

The moral right of the author has been asserted

This book is sold subject to the condition that it shall not, by way of trade or otherwise, be lent, re-sold, hired out or otherwise circulated without the publisher's prior consent in any form of binding or cover other than that in which it is published and without a similar condition including this condition being imposed on the subsequent purchaser.

ISBN 978-1-917000-04-8

To Mr and Mrs P with infinite thanks

To unlock a society, look at its untranslatable words.

Salman Rushdie *Shame*

Contents

Introduction	x
Foreword	xv
Instructions for Use	xvi
Groupings by Topic	xvi
A-Z	21
Vintage Evaluation	183
Some Dates in the Montalcino Calendar	190
Further Reading in English	191
Afterword	193
Acknowledgements	195

Introduction

One of the following statements is false: Montalcino, a tiny municipality of less than six thousand inhabitants, is a city; and I am the mayor of the town.

Montalcino is a medieval hilltop town enclosed by stone walls, within which about 2,500 of these inhabitants live. I like to think of Montalcino as an island, rising up above the rolling *Crete Senesi*, the gentle, morainic clay soils that are so typical of Tuscany. A sign upon arrival informs the visitor that we are 564 meters above sea level. The sea we perceive is an everchanging landscape of colors that one previous mayor, Ilio Raffaelli, liked to describe as a *'davanzale per l'apocalisse',* a quotation from the poet Alfonso Gatto that means 'a windowsill for the apocalypse'. It is unlikely that you will find yourself passing through Montalcino. It is more usually the case that you will want to get here. Or to be more precise, you get here only if you wish to do so. No one comes across Montalcino by chance.

Here on the hilltop, the island, things look very different compared to down on the valley floor. The word *campanilismo* was coined to define people's attachment to their bell tower or their own small community. The Tuscan people are real masters at this. I remember Pino, a great friend, mentor and *ilcinese*, ie from Montalcino. He was very serious when he made sure his daughter told people she came from a neighboring village so that if she did not do well in a sports event there would be no lasting shame for Montalcino. This community attachment is linked to the preservation rather than the irreversible consumption or exploitation of the territory. *Campanilismo* promotes identity, from the smallest of things upwards. Nowhere else than in Montalcino is the famous Tuscan pasta, handmade with just flour and water, called *pinci*, rather than *pici*.

However, a certain willingness to preserve one's identity could lead to the off-putting attitude of repelling others, foreigners, in short, the different. There is a risk of it turning into a certain guardedness and fear of contamination from the outside world, pushing back a society to within its four walls. Luckily, Montalcino, although proud of its great identity and distinctiveness, whether expressed in cultural or culinary forms, has never been difficult for people from elsewhere to access. In many cases it has functioned as a shelter. The Sienese intelligentsia are still very grateful for the fact that, five centuries ago, they had the opportunity to declare the Republic of Siena in Montalcino while escaping from the Florentines. Other less historical examples exist too. *Brunello* wine, a great local initiative consisting of privileging the Sangiovese grape in an area that had adopted the tradition of mixing black and white grapes almost everywhere, has become widely known and valued, thanks to an American winery here.

The *ilcinese* community gave another example of inclusivity when a Scottish girl, fresh from Oxford University, entered the walls of Montalcino in 1995. Laura Gray, nearly thirty years ago, can only have been even more enthusiastic, energetic and sharp than she is today. She found Montalcino recovering from its late 1970s status as Tuscany's poorest municipality. *Brunello* was really starting to gain traction and attracting large numbers of tourists, thus completely changing the face of Montalcino's community and opening it up to prolific international contamination. Laura is a humanist. She is interested in people, a listener and an acute observer who rapidly builds thoughts upon topics and develops ideas. She was probably almost too much for the community back then. With her joyful character, she catalogued the city public library which could then reopen after a long closure. Employees at that time were not overimpressed by her initiative and get go.

At some point, an enlightened employer gave Laura the opportunity to channel all her positive energy and intelligence and that is how she came to be at the helm of a Montalcino winery for nearly twenty years. She grew her family, built a brand and a cellar but, in particular, she built up knowledge and relationships. Many Italians feel that our culture is too complex, deep and intricate to be explained by a foreigner but, in my opinion, a clear and more Anglo-Saxon reading of our culture is exactly what is needed to let the world know about our story and culture. Some years before Laura's arrival, Australian-born Isabella Dusi settled in Montalcino, and her experience here culminated in a first great introduction to the town: *Vanilla Beans and Brodo* (2002). I believe that Laura's book comes just at the time we need an updated introduction, a clever way of looking at popular beliefs, tradition and idiosyncratic names and customs.

So, is Montalcino a city? The concept has evolved but I consider a city to be a complete ecosystem of people, places, culture with a certain distinctiveness, regardless of the number of the people living there. Montalcino was actually declared a city back in the 1462 by Pope Pio II and I am fine with that, and no, I am not the mayor, although I have been asked many times if I am, as I know every single human being in the community here and most of the stories connected to them and to it. I am a strong believer in this community and in its potential.

While most of the people here are relatively conservative, we have been lucky enough to have had mayors who really saw beyond the short-term problems and made sure that we can now enjoy a territory that is full of biodiversity, clean, free of factories and not mindlessly and uselessly covered in concrete. Quality of life is just on another level, thanks to crystalline air. There is a low population density; just six thousand people inhabit the 31,500 hectares that comprise the Montalcino territory: under fifty inhabitants per square mile, which is about 0.5% of the population density of London. Although vineyards are the highest value activity, there is no such thing as monoculture. The slopes are still covered in forests, fields and olive groves. The ratio of other cultivations versus vineyards is 8:1 – heaven for the eyes. We live in a beautiful, sunlit paradise, one that is far from airports, motorways and other means of transport, yet that is far from being provincial or backward-looking. On the contrary, here you can find great numbers of foreign cognoscenti, entrepreneurs, artists, nature-lovers and, of course, wine lovers.

Montalcino is the city of *Brunello* and, therefore, one of Italy's wine capitals, but it can also be a paradise for the non-drinker. Back in the days of the Roman Empire in Gaul, the Latin poet Ausonius said '*Hic uva ubique nomen*' ('Here the name of the wine is everywhere'), referring to Bordeaux's reputation around the world. Well, here we could say *Brunello* is everywhere, yet there is only one Montalcino, because when it comes to this city, being excellent is not the whole story; what also counts is being unique. Finally, thanks to our Magpie, we have a way to share this uniqueness with a larger audience.

Gabriele Gorelli MW

Foreword

One of our overused family phrases in 2023 was the incredibly rude 'Who asked?' and I must admit that I have often posed this question to myself during work on this peculiar volume. I have been down many rabbit-holes and have also been the recipient of kindnesses from unexpected sources.

The idea for this project came to me in 2022 after a huge change in my professional life. My time running a winery in Montalcino had come to its natural end. I needed to find a way to commemorate a significant chapter of my life. At the time, the vague answer of 'writing my book' served both as a smoke screen and a Trojan horse, allowing me to ask questions and make a different set of connections. The dictionary format lends itself to the kind of knowledge in which I excel. In spite of some of the negative connotations attached to magpies, what better bird than that infamous collector of shiny treasures to announce my unserious intentions? I imagine my readers hopping, bird-like, through the entries until they find one that sparkles to them.

This is my gift to Montalcino, home for a good portion of my life.

Laura Gray • Spring 2024

Instructions for Use

Do not consume in one sitting. To avoid indigestion, take small sips.

Words in bold are entries in their own right. Sometimes plurals get in the way of this system and I have opted for common sense.

Mistakes are all mine. Consistency is not a magpie strongpoint and these three hundred entries are by no means exhaustive. I have eschewed a few minefields and have chosen not to dwell on the achievements of specific wineries.

Not all the alphabetical entries feature below (there are numerous hidden treasures for magpie readers), while some appear in more than one category.

Groupings by Topic

Acronyms
AAD, ABV, AIS, DOC, DOCG, ICQRF, IGT, MOG, OTBN, UFO, VIPs

Barrels
Affinamento, Airlocks, Barrels, *Barriques*, *Botti*, *Colmatori*, *Garbellotto*, *Imbottigliamento*, Slavonian Oak, *Tonneaux*

Bottles
Bottles, *Fascette*, *Imbottigliamento*, Large Formats, Lightweighting, Neckstrips, Punt, *Sughero*, Ullage

Brunello: Rules and Regulations
AAD, ABV, *Affinamento*, Altitude, Bottles, *Consorzio del Brunello*, Denominations, Density, *Disciplinare*, DOCG, *Idonietà*, *Imbottigliamento*, Irrigation, ICQRF, Large Formats, *Resa*, Yields

Challenges
Afa, *Brett*, *Caprioli*, *Cinghiali*, Covid-19, Drought, *Etilfenoli*, *Fillossera*,

Flavescenza Dorata, Frosts, *Gelicidio*, *Grandine*, Heatwaves and Heat Spikes,
Millerandage, *Oidio*, *Peronospera*, *Quercetina*, *Stress Idrico*, Vandalism

Climate Change
ABV, *Agostamento*, Altitude, Canopy Management, Drought, Frost,
Heatwaves and Heat Spikes, *Maturazione*, Phenological Stages, Snow

Etiquette
Aeration, Decanting, Drinking Window, Glassware,
Tastevin, Vintage Evaluation

The Etruscans
Cypress, Etruria, Fufluns, *Oinochoe*, *Poggio Civitella*

Geology
Alberese, *Galestro*, Geology, *Scheletro*, Subzones, *Terroir*, Whale in a Vineyard

Gods of Wine
Bacchus, Dionysus, Fufluns, Liber

In Your Glass
Brunello, Gin, *Grappa di Montalcino*, IGT, *Moscadello di Montalcino DOC*,
Riserva, *Rosso di Montalcino DOC*, *Sant'Antimo DOC*, Super Tuscans

Italian Language and Culture
Bella Figura, *Comune*, *Enoteca*, Hectoliter, Hectare and Hectogram,
Inventory Loan, J, *Mezzadria*, Months, *Ponti*, *Quintale*, *Santo Padrono*,
Spampanamento, Suffixes, Superstition, Unification of Italy

Landscape
Bosco, *Creti Senesi*, Cypresses, Fogscapes, *La Torre dei Pomodori*, *Leccio*,
Macchia, Monte Amiata, Ombrone and Orcia, *Poggio*, *Recinzioni*, Val d'Orcia

xvii

Life in Montalcino
Apertura delle Cacce, Archery, Bicycles, *Borghetto*, F402, Hospital, Nicknames, *Pianello*, *Presepi*, *Quartieri*, *Ruga*, *Sagra del Tordo*, Santo Patrono, School, *Sorteggio*, *Travaglio*, *Trescone*

Montalcino Fauna
Caprioli, *Cinghiali*, *Recinzioni*, Ungulates, Wolves

Montalcino Flora
Albatro, *Bosco*, *Cachi*, Cypresses, *Finocchio*, *Funghi*, *Ginestra*, *Leccio*, *Macchia*, *Pungitopo*, Roses

On the Map
Camigliano, *Cappellone*, *Fortezza*, *Giardini dell'Impero*, *Il Bosco della Ragnaia*, La Madonna, *Palazzo dei Priori*, *Panfilo dell'Oca*, *Passo del Lume Spento*, *Piazza del Popolo*, *Piazza Padella*, *Il Prato*, *Poggio Civitella*, San Giovanni d'Asso, *Teatro degli Astrusi*

Personaggi
Raffaelli, Gambelli, Gatto, Gorelli, Hemingway, *La Titina*, Marinetti, Moglio, Poirot, O'Keefe, Temperini

Proverbs and Sayings
Candelora, *Contadino*, *Estate di San Martino*, *Funghi*, *I Giorni della Merla*, Suffixes, Superstition, Snow, *Zappatura*

Relationship with Siena and Florence
Assedio, *Beccamorti*, *Campanilismo*, *Gian Gastone*, *Fortezza*, Relationship with Siena, Wolf Statue

Sporting Events
Brunello Crossing, *Eroica*, *Strade Bianche*, *Giro D'Italia*, *Val d'Orcia Gravel*

Terminology: *in cantina*
AAD, *Cali*, *Cantiniere*, *Coacervo*, *Colmatori*, Concrete, *Délestage*, *Feccia*, *Fermentazione*, Fining Agents, *Enologo*, Malolactic Fermentation, Marcs, *Massa*, Must, Pump-over, *Solforosa*, *Travaso*, Winemaker, *Volatile*, *Vinaccia*

Terminology: *in vigna*
Accappanamento, *Agronomo*, Bug War, *Campionatura*, Canopy Management, *Cordone Speronato*, *Cimatura*, *Diradamento*, *Fattore*, Growth System, Guyot, Irrigation, Sexual Confusion, Yields

The Life of a Vine
Barbatella, Bud Break, *Invaiatura*, *Agostamento*, *Maturazione*, Phenological Stages, *Sfarfallamento*

Things to Eat in Montalcino
Pinci, *Corollo*, *Miele*, *Ossi di Morto*, *Tartufo*, *Zafferano*

Things to Look at, in and around Montalcino
Fiaschetteria Italiana, *Fortezza*, *Gian Gastone*, *Mattonella*, Mosaic Roundabout, *Museo della Comunità di Montalcino e del Brunello*, *Oro di Montalcino*, *Sant'Antimo*, *Temperini*, *Tempio del Brunello*, UFO

Vineyard Operations
Accappanamento, *Cimatura*, *Diradamento*, *Potatura*, *Sfemminellatura*, *Sovescio*, *Spampanamento*, *Stralciatura*, *Vendemmia*

Weather
Afa, Drought, *Gelicidio*, *Grecale*, *Grandine*, Fogscapes, Heatwaves and Heat Spikes, *Libeccio*, *Maestrale*, *Scirocco*, Snow, Rainfall, *Tramontana*, Wind

AAD

This acronym, chalked on many a blackboard hanging from **barrels** and taped to stainless steel tanks, stands for the words *Atto A Divenire*, and must be used when describing **DOCG** wines prior to their final approbation. It can loosely be translated as the transcendental En Route to Becoming, ie a trainee or wannabe *Brunello* in its unbottled state of *sfuso*. All AAD wines aspire to being *idoneo*. Their eligibility to be *Brunello* is officially recognized when they receive *idonietà*. The next step is generally *imbottigliamento*, **bottle**-aging, labelling and, finally, release into the world at large.

ABV

Alcohol By Volume is quite a touchy subject in these days of global warming and calorie counting. As per the ***disciplinare***, the minimum alcohol level for ***Brunello*** is 12.5%. To date, there is no maximum. In recent years, it has not been unknown to see 15% on labels, as increasingly hot summers put some production areas to the test. Alcohol in wine is measured in percentage as opposed to proof. In the EU there is a 0.5% tolerance for labelling convenience. This means that a wine labelled 13.5% could be very close to 13% or to 14%. If this seems shocking, bear in mind that in the US, for wines below 14%, the tolerance is 1.5%.

The alcohol level of a wine is the natural result of the **fermentation** process, as sugars in the grapes are transformed into alcohol. Alcohol affects the taste, texture and structure of a wine. If there is enough of everything else, ie if the wine is balanced, then the alcohol level may not be evident to the person drinking (so no flushed cheeks), though this does not mean you are safe to be driving. Of course, a certain level of alcohol is necessary to sustain a serious wine with a long cellar life ahead of it; alcohol, along with **acidity** and **tannins**, forms the backbone of a wine and is a powerful preservative that permits aging.

Accappanamento
Almost as good as Mississippi for a spelling competition, *accappanamento* is an alternative to *cimatura*. Embedded at the heart of the word is *cappana* meaning hut. Rather than lopping off the tops of the shoots, growers can choose to bend them over to form a natural parasol. This can be a good option in years of heat and **drought** since it does not force the plant and creates some shade. Afterwards, the *stralciatura* is trickier since the vines lignify as they are, wrapped tightly around the support wires, which makes removing them time-consuming and tedious.

Acidity
Acidity is an important component of wines and is responsible, in the best scenarios, for mouth-watering juiciness. Its absence results in flabby, dull creations. It is composed of many different acids but is usually measured by the presence of the most common: tartaric acid. The minimum total acidity for **Brunello** as per the **DOCG** regulations is 5 g/l. *Sangiovese* is a grape with elevated acidity, although hot summers increase alcohol and can decrease acidity.

Aeration
The market for wine accoutrements is saturated with aerators, decanters and belabored metaphors referring to the best methods of waking up teenagers and grandparents. Wine that has been cooped up in a **bottle** needs time to loosen and become fully expressive. As well as opening and releasing aromatics, exposure to air via **decanting** simulates time in the cellar and allows full potential and character to emerge. Aerating can help to soften **tannins** and release fruit flavors or reduce the intensity of *volatile* **acidity** and other undesirable compounds. Rather surprisingly, the hard science seems to indicate that opening bottles ahead of time actually has

minimal effect due to the diameter of the neck opening. In short, it is a free world. If your fancy aerator was a Christmas gift, you should definitely use it, but nothing beats the practice of giving time in the glass to allow a wine to evolve and tell its story.

Afa

These three letters (that apparently mean Grandpa in Icelandic) are used Italy-wide to convey the misery of hot and humid air, oppressive and enervating heat. Sultry does not cover it. The word *afa* can be exhaled as a sigh and, certainly, for vineyard management it necessitates extra vigilance for all the spore-spread diseases.

Affinamento

Affinare is the process of refinement through age and the verb is used as an alternative to *invecchiare*. In Italian, *affinare* is deemed more elegant, though this does not translate perfectly to English, where refining conjures up images of petroleum and sugar, while aging is graceful and desirable. The original **DOCG** *Brunello* rule was for three and a half years in large oak **barrels**. In 1991 this was reduced to three years. In 1996 the strictures regarding size were lifted and the rule was changed to include barrels '*di qualsiasi dimensione*', meaning of any and all sizes: basically, anarchy. The duration was reduced to two years in a 1998 ruling to be applied to vintage 1995 onwards. The change in legislation was to accommodate producers' voluntary transition from aging in large ***botti*** to using smaller ***barriques*** and ***tonneaux***. There is no rule regarding the origin of the wood, which can be from Slavonia, France, Germany, Austria, America or even Mongolia.

Agostamento

This term derives from the Italian word for August, which is, in fact, the month of the year in which the vine's primary shoot structure matures and lignifies, ie the green color disappears and it becomes woodier. Climate change is pushing this forward, alarmingly close to July. This process is powered by the transformation of sugars to starch that is exported as reserves for next year. The shoots begin to lignify as the grapes begin to ripen. This process allows the vine to survive winter and is important for the following year's production; it occurs almost simultaneously with *invaiatura*. In the vine's annual cycle, *agostomento* corresponds to the second highest peak of root activity, the first being **bud break.**

Agronomo

An agronomist, or agricultural engineer, is an expert in the science of soil management and crop production. The term derives from the Greek for laws and management. Agronomy is a field of study within the profession of agrology. Beloved by Canadians, the term agrology was coined in France and was first used in 1849. It derives from the Greek words 'agros' for land and farmer, and 'logia' for science. It is not in use in Italy and therefore we do not have *agrologhi* in Montalcino. However, we do have an upcoming generation of home-grown *agronomi*, thanks to the 2017 opening of a specialized secondary **school**.

Most wineries here will have a qualified *agronomo* working in some capacity, either as permanent staff or as a consultant. This is the person who will influence how best to manage the growing season. They might make suggestions about fertilization products and methods, how and when to prune and have input regarding grape thinning or **canopy management**. Ideally, agronomers cooperate closely with the winery **enologists**, since the best wines are made in the vineyard. In the last forty years, the significance of these two roles seems to have switched in this area of Tuscany. In the 1980s, **winemakers** were the rockstars and the professional figures in which to invest. Now, with climate change and every season presenting unprecedented weather, a good agronomer is a huge advantage and can make the difference in a difficult vintage. This is the person who will know the best powdered rocks to protect against a late **frost** or who can be relied on to suggest new strategies for managing **drought**. Keep on speed-dial in case of hail and other time-sensitive interventions.

Airlocks
The intersecting glass bulbs at the top of large **barrels** provide valuable information regarding the contents within, as well as an important clue to a winery's attention to detail and hygiene. *Colmatori*, as they are known in Italian, feature amongst a short list of items attributed to, but **not invented by Leonardo da Vinci.**

AIS
Ais, Ais Baby. AIS is the Italian Association of Sommeliers. This whole book revolves around being able to write those three words. It has existed since 1965 and has over 40,000 members. The other accredited institutions are FISAR (*Federazione Italiana Sommelier Albergatori e Ristoratori*) and ONAV (*Organizzazione Nazionale Assaggiatori Vino*).

Albatro
The *corbezzolo* or strawberry tree is known as *albatro* in Montalcino. In December, the *macchia* is dotted with the attractive red pom-poms that are the fruit of this shrub. The species name (*arbutus unedo*) is attributed to Pliny the Elder who apparently declared '*Unum Tantum Edo*' (I eat only one). This ambiguous phrase can be interpreted to mean either that they are inedible or that they are such a treat that one is more than enough. My vote is for the first interpretation. They look so much more beautiful than they taste and, should you be foolish enough to try – even just one – you will find yourself picking seeds out of your teeth for months afterwards. Honey made from the blossom is another matter entirely.

Alberese
Alberese is a sleepy coastal town, the best place to leave your car for a wonderful day in the *Parco Naturale della Maremma*, and only an hour's drive from Montalcino. *Alberese* is also the name of a soil found in Tuscany: in Montalcino, particularly in the south-eastern areas, and in the hills where *Chianti Classico* is produced. It is a weathered calcareous marl with

a high limestone content, and is 35-55 million years old. *Sangiovese* grown on *alberese* develops savory and flinty notes. The name comes from the Latin for white, *albus*, because of the high levels of calcium carbonate. Unlike *galestro*, which is fragile, *alberese* is very hard and for this reason was beloved by Tuscan builders because it is dense, compact and almost unbreakable.

Altitude

Driving into Montalcino from Siena means that you are greeted on arrival with a large, hand-lettered sign indicating the altitude of the town: 564 meters. Carry on towards Grosseto and you reach the highest point, ***Passo del Lume Spento***, 621 meters above sea level. Historically, Montalcino's position was extremely valuable and was why the town was claimed by both Florence and Siena as a strategic stronghold. The long and beautiful views that we now enjoy used to be a military advantage, and the stubby watchtowers that can be spotted all around were not designed to be future tasting rooms but had an altogether different purpose.

Until 2015, the **DOCG** disciplinary specified 600m above sea level as the maximum altitude for **Brunello** vineyards. A few wineries had vineyards that were too high and were forced to label as Chianti Colli Senesi (a DOCG since 1984), made in Montalcino from 100% *Sangiovese*. The 600m limit was removed in 2015 and as a result some great plots were admitted to the fold. Some might say that one of the effects of global warming is that quality is crawling up the slopes of Montalcino. The most valuable land is now generally considered to be '*in quota*,' benefitting from later **bud break** (helpful with late **frosts**), more extreme day/night *escursione termica* and longer hang time.

High altitude wines (grapes grown at least 500 meters above sea level) are trendy in the wine world; some restaurant wine lists even use the elevation of the vineyards as criteria for dividing the wines they pour. Higher altitude wines tend to have less overbearing alcohol and more elegance: the goal of the enlightened consumer.

Ampelography

This word refers to the study and classification of vines and grape varieties. Botanical observation and now DNA and molecular science are used to establish parentage and pedigree. The ampelographic diversity of Italy is huge and the adjective *ampelografico* is understandably popular amongst Italians. Italy grows the largest number of native grapes known, amounting to more than a quarter of the world's commercial wine types. The Italian Government defined the Italian General Ampelography in 1872, scarcely a decade after the **unification** of the country. At present, the National Register of Vine Varieties numbers 610 grape varieties for wine production and 191 table grapes. It is easy to consult; just look up *catalogoviti* online.

Apertura delle Cacce

The *Apertura delle Cacce* (the Opening of the Hunting Season) is a town festival and is the precursor to the **Sagra del Tordo**. Together, these two festivals are known as *feste identitarie* and are embedded in the Montalcino statute, insofar as they refer to the socio-cultural identity of the population. *L'Apertura* is always on the second Sunday of August. In spite of the heat, the medieval pageant takes place with a lot of velvet and chain mail, culminating with the **archery** competition between the four quarters on the playing field below the **Fortezza**.

Archery

It is no coincidence that some of the best archers in Italy come from this area of Tuscany. Montalcino offers a limited range of sports to children growing up here but archery is one of them. Each *quartiere* has its own pitch and is keen to raise a new generation of champions. The focus of the **Sagra del Tordo** and the **Apertura delle Cacce** is the archery competition on the football pitch below the **Fortezza**. The winning quarter receives a silver arrow and celebrates with as many dinners as victories. Lots are drawn for the shooting distances and the archers who will be allowed to shoot for their *quartiere*. This is barebow archery, which means that the archers shoot at targets without the assistance of a mechanical sight or stabilizers. The targets (with a wild boar silhouette) are set at 25, 30, 35 and 38-45 meters.

Assedio

From the twelfth century onwards, Montalcino was an autonomous *comune.* The town was also a contested and valued strategic outpost until the Renaissance. It was an impenetrable stronghold, protected by walls and a great fortress. Over the centuries, Montalcino faced fierce military conflict, prior to 1360 against Siena, and then, as Sienese citizens, against Florence. You can see the traces of each reign, the black and white shield of Siena (*la balzana*) or Medici balls plastered on various buildings around town. Seen from this end of history, it is hard not to assume that this elegant one-upmanship or medieval 'tagging' was a source of humiliation to the proud townspeople.

Sources from the time indicate that the ***ilcinesi*** were perfectly happy with whomever they were aligned with at the time (it was, after all, the time of Machiavelli and the Borgia family) so long as their trade privileges remained intact. Some shades of these historical animosities still exist. Apparently, the Medici balls on the *Palazzo Pubblico* in *Piazza del Campo* in Siena are never cleaned by the Sienese workers on principle and, in fact, they do look positively dingy.

But I digress: back to the *assedio*, or siege, of Montalcino. Besieging cities was all the rage in the Renaissance (donkeys were once catapulted into Siena, where a siege lasted fifteen months) and for almost three months in 1553, the Montalcinesi were besieged within the walls of the **Fortezza** by the Imperial and Medicean forces, including Spanish troops led by Don Garcia of Toledo. The invading army was reported to number over twelve thousand men. The legend is that Biagio de Montluc, the heroic defender of Sienese liberty, stood on the *Fortezza* battlements and patted wine onto his cheeks so as to hide the signs of starvation and appear in good health.

Another account from the time tells of the day before Don Garcia's troops retreated, when a Spanish guard played his guitar and the Montalcino womenfolk got up on the ramparts to sing and dance. It's hard not to extrapolate something about the character of the people and I imagine this particular scene involving a degree of mocking and skirt-lifting, rather than a musical version of the 1914 Christmas truces.

Autoctono
This is the term used to define indigenous or native grapes (and **yeasts**). In English, it is the mouth-filling 'autochthonous'. Few grapes are truly indigenous and *Sangiovese* certainly does not fall into this category. In all likelihood, this grape is native to Calabria and is the genetic parent of *Nerello Mascalese* and *Frappato*.

Babo

Not to be confused with Tuscan for daddy (*Babbo:* as always, beware the double consonant in Italian pronunciation), this is a scale for measuring the grams of sugar in the **must**. It measures the specific gravity, or density of the must and has to be adjusted for temperature. One *Grado Babo* (expressed with the symbol Kmw) corresponds to 10 grams of sugar for each kilogram of must. The inventor was the Austrian baron and **winemaker** August Wilhelm von Babo (1827-1894). Ideally, a fermenting **Brunello** should have an initial *Babo* of between 20 and 20.5 to yield the elusive 13.5% **ABV**. A *Babo* of 21 or 21.5 means 14% or 14.5% on the label. Just like **Brix**, *Babo* is an indicator of potential alcohol in the final wine.

Bacchus

Bacchus is the Roman god of wine, agriculture and fertility. He was also the protector of the theatre and patron of the arts. His festivals, the *Bacchanalia*, were celebrated with great enthusiasm and extreme revelry. In modern day Italy, *'Per Bacco!'* is a comedy expostulation, as popularized by *Topolino*, the Italian Mickey Mouse. This phrase is generally used by someone trying not to swear or blaspheme, rather like saying 'Sugar!' or 'Fudge!' And, just to make things more confusing, Bacchus is also the name of a hybrid grape created in 1933 in Germany from Silvaner, Riesling and Müller-Thurgau. Released for general cultivation in 1972, it does well in the UK, where it is considered to be the English alternative to *Sauvignon Blanc*.

Barbatella

A baby-vine, just as sweet as it sounds. It is not a coincidence that they are raised in nurseries.

Barrels

You will need to be more specific. In Montalcino, size matters. Whether a producer is an advocate of **traditional *botti***, and a longer *affinamento* (aging) in wood or a fan of smaller *barriques* or *tonneaux*, the choice of wood dimensions and origin can say a lot about the producer's goal for their wines, their ageability and even the provenance of their ideal consumer.

Barriques

Barriques are small **barrels** (225 lt) used by producers who favor a shorter, more intense wood-aging and consequently a longer **bottle**-aging before hitting shelves. The obligatory wood-aging for ***Brunello*** is two years but the wine cannot be released until the fifth January after harvest. A *barrique* contains the equivalent of three hundred bottles of wine.

Beccamorti

Being a *beccamorto* was a medieval profession. The job consisted of biting big toes as a way of ascertaining whether someone was alive or not. Banks employed *beccamorti* to check if debtors were pretending to be dead and that is why, to this day, corpse toes are tagged. Doctors were known as *beccamorti*, as were gravediggers... and so are ***ilcinesi.*** In September 1260, the Sienese Ghibelline army beat the Florentine Guelph faction at the Battle of Monteaperti. It was a brutal battle; Florence lost 30,000 infantry and 3,000 cavalry troops. Dante describes the local river (the Arbia) running red. Montalcino was allied with Siena but, legend has it, turned up intentionally late to the battlefield and risked compromising the successful outcome. As a punishment, the Montalcino soldiers were made to bury the dead and so were given this insulting **nickname**.

Bella Figura

More than *La Dolce Vita*, this idiomatic expression seems to me to be an essential aspect of life in Italy, a country of visual appreciation and attention to detail, but it verges on the untranslatable. *Bella Figura* means to cut a fine figure, but it signifies much more than the right clothes, looking good or the right time to drink a *cappuccino*. My daughter, aged eight, was once reprimanded by a British friend her own age for 'showing off'. She was utterly perplexed by a concept that she had never encountered in her Italian life to date and, even when it was explained to her, was hard put to understand what could be negative about it. Achieving or maintaining *bella figura* requires a deep understanding of Italian culture to gauge what is appropriate, whether behavioral or linguistic. When choosing a present in a shop, you might explain to the salesperson your desire to *fare bella figura*, ie make the right impression. The opposite is *fare brutta figura*, the result of gaffes or cultural misbehavior and is infinitely easier to accomplish, particularly for foreigners.

Benvenuto Brunello

An annual **kermesse** to welcome and promote the new vintage of **Brunello**, first edition 1992. It was the first *en primeur* event for wine in Italy. Originally held in February in a series of ever larger locations, in 2021 it was rescheduled to November, to better coordinate producers, journalistic impressions and market allocations before the actual release of the vintage in the following January. Never try to get your hair cut around *Benvenuto*, since all of Montalcino is sprucing themselves up for the onslaught of visitors. Many producers are unrecognizable at this time of year.

Bicycles

Bicycles are a thing in Montalcino, more so than one might expect for a hill-town surrounded by hill-towns. There are a number of cycling events punctuating the year, trapping residents in their homes for days at time while roads are closed to traffic. Montalcino's **altitude** and unpaved roads, along with the postcard beauty of the **Val d'Orcia** mean that it is an attractive destination for all kinds of cyclists; mountain bikers following the route of the now defunct **Granfondo**; apocryphal hipsters on penny farthings for the **Eroica** (perhaps not strictly true… nevertheless the costumes are marvelous and the athletic preparation required is considerable); gravel bikes for the **Strade Bianche** or **Val d'Orcia Gravel** and racing bikes retracing the steps of the *Giro d'Italia.*

You may find yourself waiting for kilometers at a time, in the hopes of overtaking a peloton of cyclists on the serpentine slopes of Montalcino. While you curse them at every curve as you experience acute fear about the abrupt arrival of a speeding car from behind, spare a moment to consider the social revolution linked to the availability of safe and relatively cheap bicycles at the end of the nineteenth century. The average distance between engaged couples went up from walking distance – say five to eight kilometers – to three or four times as much, with wonderful effects for the gene pool. Studies of British parish records show that marriages between people in the same villages almost halved between 1907 and 1916, corroborated by the decreasing of surname clusters. I cannot help but think of my father-in-law valiantly pedaling from Sant'Angelo in Colle to Cinigiano for a dance (21 km, a good two hours of incredible hills each way) in the 1950s.

Biondi Santi
Impossible not to mention the so-called founding father of **Brunello**, Ferruccio Biondi Santi, though the tale of this family, crucial to Montalcino's history, is told so well elsewhere that I will not attempt to summarize it here. A great moment for Montalcino was in 2000, when the *Biondi Santi* 1955 ***Riserva*** was the only Italian wine to feature on the list of the best ten wines of the century in *Wine Spectator*.

Borghetto
One of the Montalcino ***quartieri***, with a red and white flag. The Patron Saint, *Sant'Egidio*, is celebrated on the first Sunday of September.

Bosco
For such a prestigious winemaking area, Montalcino still has an enormous amount of woodland. Half the territory, around twelve thousand **hectares**,

is covered in thick woodlands, or *bosco*. A town council survey has counted over 2,300 hundred-year-old plants. The very name of the town comes from the holm oaks (*lecci*) that are so typical to this *macchia*. Wood was essential to Montalcino's early industries: tanning, ceramics and wrought iron. After the invention of the railway, the woods became a vital source of income. One thousand hectares were cut each year, rotating every twelve years, producing huge amounts of wood coal for both domestic and industrial uses. Wood was used to create railway sleepers, vineyard posts and sought-after pipes. At one point, the woods gave work to around 800 people, including Montalcino's beloved mayor, **Ilio Raffaelli.** In the 1950s, however, fuel markets evolved to rely on electricity and gas and the Montalcino woodsmen found themselves without employment.

Botti

Hard for some Anglophones to pronounce without a smirk, this is the name for the large **barrels** historically associated with Montalcino that contain upwards of three thousand liters (four thousand **bottles** worth). It is difficult for me to forget the journalist who excitedly proclaimed his love for big *botti* (the singular is *botte*) and spent the rest of the morning apologizing. The classic wood for *botti* is **Slavonian oak.** It is prized for being relatively neutral but the size of the *botti* is key since it affects the ratio of surface area to volume; essentially, the larger the cask, the fewer oak lactones and oxygen are imparted into the wine. Large *botti* are permeable recipients for the wine to become itself rather than active flavor contributors. After one year of use, these barrels give very little **tannin** to the next batch of wine. And yes, they are often cleaned by having a person squeeze through that small silver door, preposterous as it seems. As long as your shoulders are the widest part of your anatomy you can slither in and stand up inside.

Bottles

The ***disciplinare*** specifies that only dark glass may be used and the bottle type must be *Bordolese*. This style of bottle originated near the French city of Bordeaux and has a cylindrical shape, accentuated shoulders and short neck. Do not confuse the Bordolese with Burgundy bottles that are altogether stubbier; they are sometimes described as pear-shaped with rounded shoulders, though this seems a little too close to bottle-shaming for my liking. One theory is that these differences arose to help quick regional identification due to illiteracy and label deterioration in seventeenth-century France. The rules make it very clear that **Brunello** bottles cannot be 'dressed' in any way incompatible with the prestige of the wine. No gimmicks here, thank you very much. If you are interested in bottles, make a stop at the *Museo della Bottiglia e Del Vetro J.F. Mariani* in the atmospheric *Poggio Alle Mura* castle.

Brett

Brett is the number one enemy in the cellar and tasters exchange knowing smirks when they condemn a wine for being '*bretty*'. Microorganisms are responsible for 'making' wine during fermentation but they can also be detrimental. **Yeasts** belonging to the *Brettanomyces* family, can, under the right conditions, form molecules that are responsible for a series of aromas. When these yeasts are found in small quantities (or early enough) these aromas recall soot, smoke and burned plastic. When a wine has been strongly contaminated, they can present as tangy notes of wet fur, and gamey, barnyard smells that range from henhouse to horse sweat, or Elastoplast. *Brettanomyces* can be transmitted from the vines themselves via harvesting equipment to any surface. Fruit flies can give the yeasts a lift from surface to surface. Even the most scrupulous cellar cleaning cannot always banish the pesky things. Wines are most vulnerable during wood-aging and prior to **malolactic fermentation**. In this instance, the low levels of sulphur dioxide and presence of residual sugars create a fantastic breeding ground. Some consumers and producers actively enjoy and even seek a light note of *Brett*, considering that it lends an element of complexity and maturity if well-integrated, while others are alarmed by the first hint of phenol or animal traces. Each to his own, but unremitting hygiene and regular tasting of aging wines are the only line of defense against these ubiquitous and potentially problematic yeasts.

Brix
Nothing to do with mortar or Pink Floyd, the Brix Scale is a way of assessing sugar levels in both the vineyard and the cellar. It was invented by a German mathematician, Adolf Brix (1798-1870). In the vineyard, you use a refractometer. A grape is smeared between the refractometer's prism and light-defusing plate before being held up to the sun; the assessor squints at the evidence like a pirate with a spyglass. In the cellar, hydrometers are calibrated to show the percentage of sucrose by weight. The Brix degrees indicate the quantity of dissolved sugar in a liquid solution. One degree of Brix means that a hundred grams of liquid solution contains one gram of sugar. Thus, the higher the Brix value, the sweeter the liquid solution. High Brixes mean high alcohol, which can impede fermentation. Another measurement for sugar levels in the **must** is *Babo*.

Brunello
B is for *Brunello*, which is, in fact, the way that my children learned their alphabet at Montalcino primary **school**.

Brunello di Montalcino is Italy's first **DOCG**. The name *Brunello* was given to what was believed to be a grape variety but, in 1879, the Ampelographic Commission of the Province of Siena determined that ***Sangiovese*** and *Brunello* were the same grape variety. Hence the confusion; it is a grape, but also a wine, made by over 200 producers, all following the same rigid and prescriptive recipe, and yet no two wines are alike. In 1992, the ***Consorzio*** changed the wording from 100% *Brunello* (*Sangiovese Grosso*) to 100% *Sangiovese* (known as *Brunello* in Montalcino).

Brunello Crossing
This is a sporting event for trail runners that is held on a weekend in February. There are three routes (45km, 24km and 14km) and the first

edition was held in 2018. There is also a non-competitive 13km that includes stop-offs at wineries, together with honey and saffron tasting. Just like the ***Granfondo del Brunello***, now replaced by the ***Val d'Orcia Gravel***, this event was created by local people keen to share the joys of their territory.

Brunellogate

This scandal rocked Montalcino in 2008. It marked a Darwinian moment for the denomination and an increase of attention to process and provenance across the board. The vintage in question was 2003, a year renowned for very high temperatures. Since Montalcino has spent the past fifteen years trying to distance itself from this sorry episode, I am afraid you will have to do your own research if you want to know more. There is a historical precedent, though; in 1472, Montalcino's bishop was similarly accused of having tampered with the local wine.

Bud Break

Alternatively known as bud burst, bud break is the moment that precedes *sfarfallamento*, when the fuzzy nubs of the new vine growth 'break' to show the first furl of their leaves. It is the first sighting of the new vintage and the beginning of the annual rollercoaster for producers. Bud break is a crucial stage in the **phenological** development of a vine and in Montalcino takes place in March and sometimes April, depending on a vineyard's position, **altitude** and the **subzone**. Atypical warm weather in late winter or early spring can cause bud break to occur earlier. When temperatures are low, vines are dormant. During spring, increasing temperatures trigger the vines to wake. Water and stored nutrients – *linfa* – begin to move in the vineyard. The vine weeps. The hard nodes turn into buds.

After bud break, the vine uses the last of its stored carbohydrates to push out green leaves. The leaves multiply, deepen in color and photosynthesis begins. In the annual cycle of a vine, bud break corresponds to the highest moment of root activity.

Bug War

Instead of using chemicals, wineries can choose to introduce indigenous insects in order to restore balance in the vineyard's ecosystem. If, for example, you find yourself with an invasion of yellow or red spider mites, it turns out that you can purchase online small, teabag-like envelopes that contain tiny, beige arthropods, invisible to the human eye. They are natural predators to the dastardly mites and in three years or so, the natural equilibrium should be restored. 'It might be organic, Mum, but is it humane?' my son asked me, and I still don't know the answer. From a spider mite's point of view, is it better to go out in a chemical flash or be dismembered and eaten by another species?

Cachi

The *diospero, cachi* or *kaki (Diospyros kaki)* is a fruit tree, a persimmon, that originated in East Asia, and is one of the oldest fruit trees cultivated by man. Persimmon trees can be found here and there all over Montalcino. The fruit is ripe for picking in November and can be matured in crates. This is not an optional step; no one will forget the horror of an unripe *caco*. Keep until slushy inside and eat with a spoon. Forgo any hope of slicing and adding to salads. The trees are truly amazing, hung with bright orange globes and no leaves. They look like early Christmas decorations.

Cali

To steal the whisky terminology, this is the Angel's Share, ie the amount of wine that is lost during the wood aging process due to evaporation and absorption by wood **barrels**. It is calculated at 3%, though, depending on the barrel size and time spent in wood, it can be much more. Before requesting **DOCG** approval, **Brunello** needs to be '*calato*', ie to have lost its '*cali*'. This 3% is part of the *resa* calculations that specify that the quantity of juice authorized to become *Brunello* cannot exceed 68% of the weight of the picked grapes

Camigliano

The sleepy village of Camigliano, some seven miles from Montalcino, has a stable population closer to twenty than to thirty. For nearly fifty years, it has held the *Sagra del Galletto* on the first Sunday of October: costumes, a medieval discus-rolling competition, dancing and eating. In the late 1970s, the *Discoteca Il Pozzo* lurched into existence and most people growing up within a thirty-mile radius of Montalcino attended it at some point. I can remember dancing in my father's arms here when my legs were not yet long enough to touch the ground. Some people maintain that it is where breakdancing started and certainly many a relationship began here. It is where I met my husband. To begin with, a *'pista'* (a dancefloor) was created from donated bathroom and kitchen tiles, giving a literal sense to crazy paving.

For almost thirty years, every Saturday night from July to September, this small hamlet on the way to nowhere was the scene of an unbelievable influx of local youth who came for summer dancing under the stars. Few clubs can boast clientele ranging from babes in prams to feisty octogenarians, but this was the Camigliano demographic, along with a smattering of imported **VIPs** holidaying in Tuscany. Unsmiling elders elegantly waltzed away the first hour before the music changed and they retreated to sit, cross their arms and tut in unison. Children raced wildly under the lights until they collapsed on mattresses and their parents took to the floor. Nepo-babies and tractor-drivers were to be seen gyrating into the early hours, before the mass exodus at the end of the evening.

Campanilismo

This word derives from the word *campanile*, which means bell-tower, and describes an immoderate attachment to one's own home town and traditions. It dates from the **Unification of Italy** (1861) and was originally a pejorative term, since it described a parochialism which was inauspicious for the nascent country of Italy. During the preceding centuries, Italy had been a fragmented collection of warring city states and republics.

In spite of the Roman legacy of roads, the transport infrastructure in Italy is still underdeveloped, and for a huge period of history, many hill-town residents had no choice but to stay close to home. Factor in the ongoing national preference to remain close to family and you can see that a sense of identity that derives from where one is born is viable in a way that is specific to Italy. In countries where geographical relocation for work and study has been normalized for generations, *campanilismo* is not an option. It is not always negative and, in fact, localism and a resistance to globalization is more desirable now than it was in 1861. When it comes to food, it is precisely this that has meant that, in Italy, discrete culinary traditions have been preserved, differing from region to region, province to province, and even village to village. My mother-in-law, from Sant'Angelo in Colle, was convinced that the people from Montalcino spoke differently. They are separated by a mere ten kilometers but from a linguistic point of view, she may well have been right.

Campionatura

Campionatura, or sampling, is the act of selecting grapes from different compass points of the vineyard and from different parts of the bunch for analysis, to assess whether the parameters are present to organize harvest.

Candelora

The second of February is Candlemas Day, *Candelora*, the midpoint of winter, half-way between the shortest day and the spring equinox. It has evolved as a Christian festival but in pre-Christian times was known as the Feast of Lights, celebrating the increased power of the sun as winter gives way to spring. Here people mutter: *'Candelora, se nevica o se plora dell'inverno sèmo fuora. Ma se è sole o solicello, sèmo ancora a mezzo inverno*'. Essentially, if 2 February is cold or rainy, then the winter is over. If it is sunny, then more cold weather can be expected; basically, a version of *Groundhog Day*, based on the weather, rather than the performance and inclinations of a rodent. I was beyond shocked to discover that here, 2 February is the official day for dismantling Christmas trees and nativity scenes. This is not a tradition I want to even contemplate, much less adopt.

Canopy Management

This refers to a whole portfolio of techniques used to manage the canopy (the leaves, shoots and fruit) from pruning to harvest. It includes the initial **growth system**, green harvest and so on. Processes such as photosynthesis, respiration and transpiration are all affected by the quantity of vegetation and the exposure of the vine's leaves and fruit to sun. Care and attention to the leafy wall can involve topping (*cimatura*), or leaf stripping for dappled light. Too many leaves prevent air circulation and foster humidity which creates ideal conditions for disease.

Cantiniere

This is the person who is in charge of cellar operations. Their daily world is circumscribed by the walls of the *cantina* and their busiest time is **fermentation** post-harvest. A good *cantiniere* may be in direct contact with the consulting ***enologo*** and the ***agronomo*** and can be key to decision-making for the winery.

Cappellone

This is the name used locally for the large, vaulted space in the main square of Montalcino. There was originally a chapel (*cappella*) here, hence the name and **suffix**. The far wall used to be an open loggia overlooking the **Val d'Orcia**. Only imagine the views! It was walled up in 1958 to house Montalcino's **Post Office,** which then re-located in the 1990s to its current, massively inconvenient, position. I have read that the *loggiato* was closed by popular request since it acted as conduit for strong **winds**. The first protests were in 1914. In 2022, an *affaccio* was opened, overlooking the *Val d'Orcia*, framing San Francesco (the deconsecrated church next to the **hospital**) to perfection. The next phase is to open the arched door and other window to restore the *Loggiato del Sansovino* to its former (windy) glory. Sansovino was a Florentine architect who worked primarily in Venice and did not actually design this *loggiato*.

Caprioli

Deer are a menace during **bud break**. The precious buds are known as *gemme* – gems – and are a great delicacy to deer. The damage caused by nibbling the vines can set back production by as much as two years.

Cimatura

Topping is the viticultural practice of cutting the tips of the vine-shoots, transforming a shaggy vine to a manicured plant. This redirects the plant's energy to the remaining shoots and the fruit, and can be performed once or twice in July/August. This is an optional intervention and is not always advisable during **drought**. It can be done either by hand or by machine. The frequency and timing of *cimatura* affect hours of direct sunlight, the incidence of new leaves and the *feminelle* (lateral secondary offshoots).

Cinghiali

Wild boar, attracted by the increasing sugar levels, are a real issue as soon as the grapes turn red in August. There are thought to be five hundred thousand boar in Tuscany, forty times (!!!) the national average, concentrated principally in the province of Siena: an estimated twenty per **hectare**. In 2021, wild boar caused an accident every forty-one hours in Italy.

You can stipulate a special **ungulates** policy for car **insurance**, but the only way to protect the vines is fencing and constant vigilance. Keep eating those *pappardelle al cinghiale* and take a moment to look up *cinghiali a Montalcino* on YouTube to see an entire family of wild boar skittering down the main streets of the town.

City

Montalcino's first town council statute dates from 1415. On 15 February 1462, Pope Pius II elevated Montalcino to city status and this was ratified in 2022 by Italy's President: a superb example – or perhaps a cautionary tale – of the speed of Italy's bureaucratic systems. Entering Montalcino, you will see from the signage that it is a multi-faceted city, specializing in oil, honey, organic farming and wine.

Clones

Sangiovese goes by many names and has massive clonal variation. The **Biondi Santi** BBS11 clone was registered in 1978. The producer Banfi undertook groundbreaking research in the 1980s, and discovered about 650 different clones of *Sangiovese* on the property. After much research and a twelve-year project, they whittled them down to six, now registered as part of the national database of forty-five official clones. Other wineries have followed suit; *Col d'Orcia* has also registered three clones, as has *Fattoria dei Barbi*. Out of Montalcino, *Barone Ricasoli* and *Caprai* have registered and named clones in Chianti and Montefalco respectively. Smaller wineries tend to buy clones from nurseries who invest and research clone selection.

Coacervo

This is a fancy name for a blend of wines from the same denomination and vintage, also known as a ***massa***. This can occur when wine is moved from one barrel to another and, owing to volume differences, the leftovers need to be combined. *Coacervo* can happen by necessity or choice.

Colmatori

Barrels of any dimension should be kept full to overbrimming, in order to avoid oxidation. The Italian verb to fill, or top up, is 'colmare', and thus **airlocks** are called 'colmatori'. The airlock consists of two interlocking chambers. Water in the airlock is changed regularly and provides a barrier between the wine and the outside world. A properly maintained airlock will keep away all dangerous bacteria and microorganisms. It is also a way of keeping tabs on the wine in the barrel – without having to climb a ladder or look inside – since changes in air pressure or temperature can cause the wine to expand or contract in volume and this is easily verified from ground level. There are screeds of references online that attribute the invention of airlocks to Leonardo da Vinci and even to Galileo. Actually, airlocks were invented at the beginning of the nineteenth century. Between 1827 and 1828, various intellectuals belonging to the *Accademia dei Georgofili* debated possible improvements in wine production in conjunction with recent developments in physics and chemistry. They came up with the idea, though the first airlocks were primarily used during **fermentation.**

Coltura Promiscua

This is the opposite of monoculture, and possibly sounds more intriguing than it is. It refers to the practice of cultivating various crops in the same field: vines, olives, fruit trees, even grains and vegetables. It was once common to grow vines using fruit trees for support; they were known as *vite maritate* (married vines). In Montalcino, there are still a few vine-rows with artichoke plants at the end, but it is rare nowadays to find vine rows alternating with lines of olive trees as was once common. The system of **mezzadria** encouraged *coltura promiscua* since each family had a small plot of land that they used for all their household needs.

Commerciante

This refers to someone who deals in batches of ready-made wine, as opposed to making it from their own cultivated grapes. If you ever wondered about some of those labels only seen in airport wine shops that do not correspond to any named winery, the answer is that they may be made by a commercial operation. Some larger wineries have a commercial branch within the company and issue their second labels under that name. Sleuths may enjoy trying to trace the **ICQRF** code back to the mother winery.

Commercianti in Montalcino are regarded with none of the glamour of the French *negotiants*, historically arbiters of quality, who these days might own neither vineyards nor cellars but are the identifiable and accredited owners of the outcome. If anything, Montalcino growers feel a little superior to *commercianti*, who often stay firmly behind the scenes as if they were embarrassed to not have a true earthy connection to the grapes they buy.

Comune

Italy is divided into Regions (eg Tuscany), which contain Provinces (eg Siena, Grosseto, Firenze, Arezzo and so on). Provinces are further subdivided into *comuni* (Montalcino): municipalities or townships. *Comuni* are composed of *frazioni* (Sant'Angelo in Colle, for example).

Concrete

Wine, like most things, likes to breathe. Unlined concrete, though neutral, is semi-porous substance and allows for micro-oxygenation without imparting oak aromas or flavors. There is a growing interest in non-reactive vessels these days. As a result, egg-shaped vats are appearing in cellars all over Montalcino, for the maceration, fermentation, and aging of wine.

In the 1950s and 1960s, concrete tanks were common in Italian wineries but were generally square and vitrified, ie treated with resin which prevented the exchange of oxygen. It should be noted that although the terms cement and concrete are often used interchangeably in Italy, cement is an ingredient of concrete.

Consorzio del Brunello

I once came across this translated as the Producers' Alliance, and was rather taken by the *Star Wars* overtones. On 28 March 1966, **Brunello** was recognized as one of the first **DOC** wines in Italy. Just over a year later, on 28 April 1967, the *Consorzio del Brunello* was founded by twenty-five producers, in order to safeguard production. The signed declaration is on display in the *Consorzio* headquarters, which are on Via Boldrini, the steep road that links *Piazza Padella* with the upper levels of Montalcino.

The *Consorzio* coordinates promotional events, including the prestigious **Benvenuto Brunello** and **Red Montalcino**, collects samples for journalists, and generally works on behalf of the denomination. It is an association for producers, but membership is by no means obligatory. In fact, a growing number of rebel producers choose not to subscribe.

The president, three vice presidents and fifteen counsellors are elected from the ranks of producers, and it is quite rare for there to be more than one woman amongst this list. The 2015 *mattonella* or vintage tile commemorates the fiftieth anniversary of the *Consorzio*. Bear in mind that the *Consorzio del Chianti Classico* celebrated its first century in 2024.

Contadino

Being a *contadino* used to be pejorative. In France, the term *paysan* contains none of the implicit insult of *contadino*, a peasant or country bumpkin who works on the land and has a relatively low level of education and social status. The proverb *al contadino non far sapere, quanto è buono il cacio e le pere* can be roughly translated as 'Do not let the farmer know, how well together cheese and pears go'. Historically, cheese was a staple of the poorer classes whereas fruit was more expensive and consumed by nobles. The proverb was meant to imply that a farmer should be kept from the knowledge of how well these two products paired so that they would continue to provide them at a good price to the nobility, ignorant of their possible combination.

In November 2017, the *Fabrica Italiana Contadina*, more commonly known as *Fico* (or the Disneyland of Food), opened in Bologna. Their choice to use *Contadina* with a capital C is an indicator how working closely with the land is beginning to be considered as a source of dignity and pride, particularly for younger generations. Montalcino has a strong history of *contadino* folklore and culture.

Cordone Speronato

Cordone Speronato is the most common **growth system** in Montalcino. It is either mono or bilateral so looks like an inverted L or a T. The cordon is a permanent branch of the vine trunk that is two years or older and that is bent along the support wire. The spurs – '*speroni*' – are the growth shoots from the previous vintage.

Corollo

In Montalcino, these traditional, soft Lenten cakes are only available in the forty days before Easter. They are leavened buns made with sultanas and aniseed and should be eaten on the day they were baked. The plural is *Corolli*, and you need to know this; you will want more than one.

Covid-19

During lockdown, Tuscany, like the rest of Italy, was subject to draconian legislation that prohibited movement and promoted drone enforcement. However, agriculture was exempt and allowed to continue operations. Montalcino was empty of visitors and it was terrifically tough for the many businesses that rely on tourism, but there was no impact on the necessary and time-sensitive work that needed to take place in the vineyards. If anything, having plots in multiple **subzones** allowed workers to operate in splendid isolation. As we all know, nature was reassuringly oblivious to the pandemic and especially beautiful when left to its own devices.

In Montalcino, we are used to exercising patience, and above all, Janus-like, we are constantly looking backwards and forwards in five-year blocks; considering both the vintage on the vines that will be commercially available five years hence, and the vintage on the shelves that we need to market. During the vicissitudes of 2020 it was reassuring to think of 2025 and to be sure that when the 2020 ***Brunello*** will be released upon the world, the most surreal of times would inevitably be relegated to the status of a memory.

Crete Senesi

This is the name given to the beautiful lunar landscapes around Asciano, Buonconvento and Montalcino. The word comes from *creta*, meaning clay and the *Crete Senesi* contain great stretches of undulating grey land, punctuated by the odd **cypress** or farmhouse. It is the stuff of postcards.

Cru

Bless you! *Cru* is the French term for a single vineyard wine. The Italian term is ***Vigna.***

Cypress

Cypresses are the trees that, more than any other, characterize the Tuscan landscape, both in reality and in the collective imagination. Originally from Iran, cypresses appear to have been introduced into Tuscany by the

Etruscans and have been growing here for over three thousand years. In other parts of Italy, these evergreens are most commonly found in the proximity of cemeteries and tombs, symbolizing immortality and looming like 'softly-swaying pillars of dark flame.' (*Cypresses* by D.H. Lawrence, 1835-1930). Here, they line the roads that lead to farmhouses, punctuating skylines, and are sometimes pleasingly alternated with umbrella pines, a version of Morse code when seen from a distance. Cypresses can grow as high as thirty-five meters (one hundred and fifteen feet) and live to over a thousand years old. Look again at **Sant'Antimo** and wonder which came first, the bell-tower or the cypress? Be sure to slow down on the road to San Quirico, where there is a very famous and iconic clump, the so-called *Cipressini*, regularly described as the most photographed trees in the world and, thanks to that, an accident waiting to happen.

Many wineries in Montalcino have spectacular cypress-lined driveways. Iris Origo (1902-1988), English author and biographer, who lived most of her adult life in Tuscany, had the hillside opposite her home planted with a zig-zag of cypresses so that she could behold a Renaissance landscape from the comfort of her garden at *La Foce*.

Things I have read that I am not sure I believe:
1. Noah's ark was made of cypress wood, as was the Cross
2. The reason cypresses are popular in Tuscany is that they are very attractive to birds as nesting spots and were used as traps when birdlime was permitted

Decanting

To decant or not to decant, that is the question. Along with other highly contentious issues in Montalcino – **barrel** size, **subzones** and so on – there is a split between advocates of decanting and those who prefer the so-called 'conversation in the glass'. In spite of the abundance of beautifully shaped decanters on the market and their claims, many *Brunello* lovers prefer not to decant but enjoy the evolution of the wine when poured and, if anything, open the **bottle** for four hours beforehand. The purpose of decanting is to aerate the wine so its aromas and flavors become more expressive and to separate the sediments that may have formed.

Decanters are useful as accelerants if there is no time to allow the wine to breathe, ie in restaurant situations or in the event of surprise guests. With older vintages, the process of decanting can be necessary to remove sediment but can also be detrimental and accelerate oxidation. If a wine has sediment, the bottle should be kept vertical for twenty-four hours before opening to allow it to settle and a source of light should be used to identify the sediment. In a perfect world, if you taste a *Brunello* in its optimal **drinking window**, a glass, a corkscrew and some time is all you need.

Decisions

Gabriele Gorelli MW estimates that, in all, 1,500 decisions are made from site selection to bottling. He has consciously adapted Stephen Skelton's observation from his book, *Viticulture* (2017), a beloved text for many students of wine: 'Someone once worked out that there are 3,000 separate decisions that are taken from site selection to **bottle**, any one of which could change the character and quality of the wine in question.' This theory seemed a fascinating way to convey the constant thought that goes into making wine. Rather like parenting a baby, producers are perpetually weighing up choices, second-guessing patterns and reactions, trying to predict and offset events. In Montalcino, due to the restrictions and obligations of the **DOCG**, where many decisions are made for us, the number of choices is halved. What and where we plant is predetermined (***Sangiovese***, within the confines of Montalcino). Yield in grapes and maximum conversion to juice is similarly specified and on it goes, right down to the kind of closure, minimum **acidity** and bottle formats.

Délestage

Délestage is a two-step 'rack-and-return' process in which fermenting juice is separated from grape solids by racking and then returned to the fermenting vat to bathe the solids. Racking simply means moving wine from one vessel to another; in Italian ***travaso***. Racking the fermenting juice aerates the wine, softens astringent **tannins** through oxidation and stabilizes color. Following racking, the grape solids are allowed to settle separately from the fermenting wine. The fermenting wine is returned to the vat over the cap using a gentle, high-volume pump to completely soak the grape solids for maximum color and flavor extraction while minimizing the extraction of harsh phenols.

Demography

Montalcino in the 1980s – just as the **DOCG** was categorized – was a deprived area, receiving tax relief. How times change; the most recent studies estimate the value of vineyards to have increased by 1962% in the period from 1992 to 2022. In the 2021 census, Montalcino's population of 5,763 residents included 16% *stranieri*, ie non-Italians, coming from a total of sixty-eight different countries. It is worth remembering that in 1900 Montalcino was the third most populous town in Southern Tuscany, after Siena and Arezzo. In 1964, the motorway, the *Autostrada del Sole*, isolated Montalcino and the population decreased by 70%. The people who left Montalcino were the tradespeople, the most educated section of the population. This diaspora has had long lasting consequences, not least the abrupt break between Montalcino and the majority of its history-keepers and story-tellers.

Denominations

In 1963, over fifty years after the French had created legislation to categorize their wines, the Italian government did the same. From then on, Italian wines were divided into different denominations by the acronyms **DOC** and **DOCG**, later joined by the wild card **IGT**, forming a pyramid with the DOCG wines at the pinnacle. This was already sufficiently complicated, but the hierarchy now has three overlapping tiers as per European legislation: PDO or DOP (Protected Designation of Origin; this includes DOC and DOCG wines), IGP (*Indicazione Geografica Protetta*; this includes *Indicazione Geografica Tipica*) and plain old *Vino*, with a color specification, which encompasses the old VdT (*Vino da Tavola*) category. My suggestion is to stick to the Italian pyramid.

Density

Since 2016, all new vineyards in Montalcino must have at least four thousand vines planted per **hectare**. However, the quantity of grapes per vineyard is limited as per **DOCG** regulations. Some producers choose high density vineyards as a way of forcing quality by increasing competition for resources amongst the vines. The maximum volume of grapes that can be harvested remains unvaried, but there is more choice when it comes to selecting the very best grapes during the growing season. Moreover, the vines have to strive more to access water and minerals, and young vines are forced to extend their root system faster and deeper. As a result, the fruit is improved. Tough love works.

Dionysus

This Greek precursor to **Bacchus** is also the name of the **Mosaic Roundabout** artwork that greets all visitors to Montalcino, and is also the subject of the 2022 **mattonella**. The German philosopher, Nietzsche, considered the Greek gods Dionysus and Apollo as the exemplification of two fundamental forces of human nature: irrationality and chaos versus order and reason. On market days in Montalcino (Fridays) and during the *Sagra del Tordo*, I am sure the traffic wardens are aware of this dichotomy.

Diradamento

Bunch-thinning in order to improve and increase ventilation and fruit health is known as *diradamento*. It is performed exclusively by hand, mostly after *invaiatura*, in order to ensure the remaining bunches receive more nutrients and sugars from the roots.

Disciplinare

This is the rulebook that governs the production of **Brunello** and breaks down every aspect of **DOCG** compliance, currently standing at 43 entries. It can be downloaded from the **Consorzio del Brunello** website.

DOC

The only time a DOC has precedence over a **DOCG** is when it comes to alphabetical order. To be fair, to become a DOCG, the highest echelon of the Italian wine pyramid, a wine must have been a DOC for at least ten years. **Brunello** was a DOC from 1966 until 1980. Current Montalcino DOC are **Rosso di Montalcino, Sant'Antimo** and **Moscadello di Montalcino**.

DOCG

I quite like it when this is pronounced as two separate words, like a rapper's name; the DOC-Gee. However you choose to say it, DOCG stands for *Denominazione di Origine Controllata e Garantita*. The G in the abbreviation is fundamental insofar as *Garantita* means Guaranteed. This is what distinguishes the highest appellation of Italian wines from their lowly G-less siblings, the mere **DOC**. What the G means is that a series of stringent parameters have been both observed and controlled by an external body, ***Valoritalia***. These parameters refer to **yields**, levels of alcohol and **acidity**, **bottle** dimensions and much more as per the ***disciplinare***, which is specific to ***Brunello*** and includes both lab analyses and a panel test. *Brunello* became the first DOCG wine in Italy in 1980, thanks to a government edict (DPR 01/07/1980) that also elevated *Vino Nobile di Montepulciano* to the same status. Although *Vino Nobile*, thanks to its shorter aging time and consequent earlier release, had the honor of receiving the very first **neckstrip** (AA 000001), *Brunello* was technically the first DOCG, since it pre-dated Nobile when they were both DOCs by a whole 106 days. Pipped to the post! At present, there are seventy-eight DOCG wines in all of Italy. The highest concentrations are in three regions: nineteen in Piedmont, fourteen in Veneto and eleven in Tuscany.

Drinking Window
The best homes all have at least one of these... no, not really. The drinking window refers to the ideal period of consumption for a ***Brunello*** and can depend on the vintage and/or on the producer.

Drought
Translated in Italian as *siccità*, literally dryness, drought is responsible for causing ***stress idrico*** to the poor old vines and producers. In recent years, this has become ever more an issue in Montalcino. In 2022, under 10mm of rain fell in the months of May and June. **Irrigation** is not always a viable solution and, historically, was not permitted in Montalcino. Soil texture is very important during periods of drought.

Enologo

So much less of an offensive mouthful than oenologist, an *enologo* is the person who makes decisions about the wines during their time in the cellar from *fermentazione* to bottling and release, in tandem with the winery owners and often the *cantiniere*. Bear in mind that for **Brunello**, each vintage spends four years maturing. During these years, the **AAD** wine has to be tasted and may be moved from barrel to barrel before it is bottled. Timing is everything and regular tasting in the cellar has to happen. Is it correct to translate *enologo* as **winemaker**? I am never sure if these concepts travel across cultures with their integrity intact. Certainly, in Montalcino, there are wineries with their own resident *enologo*, either as a member of staff or in the person of the owner. That said, very many wineries share the services of just a handful of professionals. Quite often, it is not apparent from the website and materials of the producer who the consulting *enologo* is, and visitors would be hard put to 'meet the winemaker'.

Enoteca

In Italian, the **suffix** *'teca'* is used to form nouns to describe a specialization and has nothing to do with technology. The noun *'teca'* means shrine. Thus, we have *discoteca*, *biblioteca* (library), *paninoteca* (sandwich bar), *pinacoteca* (picture gallery) *emeroteca* (newspaper library), *ludoteca* (play area or games store) and, of course, *enoteca* (bottle shop) and *vinoteca* (a cross between a wine shop and a wine bar).

Eroica

This is a classic – and oh-so-marketable – *ciclostorico* **bicycle** race, where bikes and riders have to respect certain criteria. Clothing and accessories, including water bottles, must be pre-1980s and the bicycles themselves must be pre-1987 racing bikes. Cue woolen jerseys, long-socks, moustaches and irrepressible jollity. Montalcino has had its own version since 2017. The original *Eroica* was conceived in Gaiole, in Chianti, in 1997 and started a lifestyle movement. There are now versions all over the world. As of 2017, there are also Nova editions, that allow modern bicycles and gear.

Escursione termica

Also known as the diurnal shift, thermal excursion is the difference between average highest day and lowest night temperatures. In the last weeks before harvest, cool nights and hot days are the perfect combination for **organoleptic** ripening of the grapes. When temperatures drop below 20°C the skins store more antioxidants, **tannins** and resveratrol. A good September can transform an indifferent vintage for *Sangiovese* in Montalcino. What happens is that sugar accumulation in the ripening grapes slows down overnight, the vines rest and the grapes maintain higher levels of **acidity**, a key component for balance in wine. Ultimately, thermal excursion lengthens the growing season and gives the grapes more hang-time before harvest. The optimum for *Sangiovese* is 23-24°C as a maximum during the day and 12°C at night. Lower night temperatures can lead to saggy skins: heaven forbid.

Estate di San Martino

St. Martin is the patron saint of travelers and farmers, soldiers, pilgrims, innkeepers, drunks and cuckolds, and he is celebrated on 11 November, coinciding often with a few poignant mild and sunny days before winter settles in. *Estate* means summer in Italian. The proverb states: *'L'Estate di San Martino dura tre giorni e un pochino.'* This is the day Marco and I got married in Sant'Angelo in Colle back in 2000, and we were indeed blessed with three clement days. The Scottish contingent had the tops down on their convertibles, with bright clothes and bare legs (so many kilts!) whereas guests on the Italian side of the church were clad in furs.

Etilfenoli
This is a word you never want to hear in conjunction with fine wines. These are the components that signal the presence of **Brett.**

Etruria
Etruria is the name of the ancient country belonging to the Etruscans. Its area is delimited by the Arno and Tiber rivers and it roughly corresponds to Tuscany and part of Umbria. The Etruscan civilization emerged from the eighth century BCE and dominated Italy until it succumbed to the expanding Rome. In 90 BCE, Etruscans were granted Roman citizenship and in 27 BCE, the whole Etruscan territory was incorporated into the new Roman Empire.

There were consistent Etruscan settlements around the Montalcino area, principally the military garrison at **Poggio Civitella** and *Poggio alle Mura*. Enjoy the locally found Etruscan exhibits in the dedicated underground section of the Montalcino museum, and, if you have time, explore the museum at Chiusi where you can see winemaking equipment, still containing ancient grape-seeds, excavated at *Poggio Baccherina* in Chianciano. Archaeological evidence suggests that the Etruscans were trading and transporting wine. They used a two-handled beaker to drink it. They diluted wine with water, hence the dimensions of their drinking vessels. They also mixed it with cheese but we are not going to go into that.

Exposition
More usually translated as 'aspect' in English, exposition is a big deal in Montalcino, where wineries are scattered on all four quadrants of a hill at **altitudes** spanning four hundred meters. A vineyard's aspect refers to the compass direction that the slope faces (eg east, southeast, etc). Aspect affects the angle that the sunlight hits the vineyard, its total heat balance and hours of sunlight. These are all crucial concerns for **winemakers** and for the *agronomi* before them, when planning the plantation of a new site.

F402

All residents of Italy are assigned a *codice fiscale*, a sixteen-character alphanumeric code, that is required on any formal documentation. Each segment of the code refers to different aspects of identity: surname, name, year of birth, date of birth, gender and place of birth.

Montalcino is identified as F402 and only those who were born in the Montalcino **hospital**, now sadly stripped of many of its functions, have a *codice fiscale* that ends with this precious reference. Not many people are born in Montalcino these days and, in any case, since the fusion with **San Giovanni d'Asso** the code is now M378. At one point, *ilcinesi* had T-shirts made to identify the 'true' residents, a rather ill-fated idea in my opinion, though one hundred and thirty were sold.

Fascette

This word, the diminutive of *fascià*, derives from the Latin to wrap or to bind. The *fascette* or *contrassegni* are the **neckstrips** that certify that a ***Brunello*** (or any other DO wine) has been produced according to the disciplinary. They are numbered, vintage- and estate-specific, and currently show a QR code. They are usually ordered and purchased from the ***Consorzio del Brunello***, though they are produced by and can be bought directly from the *Istituto Poligrafico e Zecca dello Stato* (IPZS) – the first Italian website I have ever encountered with a Whistleblowing tab on the menu. Neckstrips certify authenticity and contain anti-counterfeit mechanisms. A *Brunello* is not a *Brunello* without a neckstrip. The current cost is just under €0.02 per strip. The demand and supply from and to producers is one way in which the *Consorzio* assesses the speed of sales from one vintage to the next.

Fattore

From the Italian verb *fare*, to do, the *fattore* is the person in charge of outdoor operations.

Feccia

In Italian, this word is used to describe the scum or the dregs of society. The winemaking term in English is much more neutral. *Feccia*, or lees, are the mix of grape skins, seeds, stems, dead **yeasts** and tartrates that deposit after fermentation and that should be discarded. Lees are a byproduct of the winemaking process that are racked off and discarded. In Italy, there are multiple obligations concerning the correct disposal of lees (and **marcs** for that matter). Wineries have the option of burying the *feccia* (with a series of permissions and controls in place) or of selling them to industrial distilleries whence pure alcohol is distilled.

Fermentazione

Since you are reading this book, I will assume that you know grapes are fermented to become wine. Producers can make important decisions about the type of **yeasts** employed and whether this process takes place in or out of wood **barrels**. Recently some producers in Montalcino have been experimenting with whole cluster fermentation, ie adding a few intact bunches, stems and all, to the fermenting juice. This has the potential to provide increased complexity, fruitiness and strong **tannins** to the wine.

Festival Internazionale dell'Attore

In the summer of 1980, Montalcino hosted the first *Festival Internazionale dell'Attore*, a brain-child of Florentine theatre producer Paolo Coccheri. Over four hundred students from all over the world stayed in the town for two months, to be mentored and taught by world experts.

In the three festivals held in Montalcino, the town welcomed first-class performers including Ingmar Bergman, Marcello Mastroianni and Monica Vitti. In 1983, the festival moved to Florence. It is currently held in Naples.

Fiaschetteria Italiana
La Fiaschetteria is a historical bar on the main street of Montalcino. It was founded in 1888 by Ferruccio **Biondi Santi**, considered by many as the inventor of ***Brunello***. It is a gorgeous example of Italian art deco, with sumptuous red velvet sofas, framed age-speckled mirrors, chandeliers and streaky yellow marble tables. It is officially a *Locale Storico d'Italia*, listed amongst around 200 other enterprises that have over seventy years of history, sport original decor and can be considered to be players in the history of Italy. The *Fiaschetteria* is known as the 'Florian di Montalcino', the Tuscan version of the wonderful Venetian café, and can count amongst its guests King Charles III, Keith Richards and, recently, Michael Douglas and Catherine Zeta-Jones. The other two *Locali Storici d'Italia* in the province of Siena are *Caffè Poliziano* in Montepulciano and *Ristorante Al Mangia* in Siena.

Fillossera
Phylloxera is a devastating louse that reached Europe in the second half of the nineteenth century. It was inadvertently brought to the continent by botanists importing American vines. Ironically, European **winemakers**, starting in France, had to graft their vines onto disease resistant American rootstocks to escape the scourge. The tale is told in a wonderful book by Christy Campbell, *Phylloxera – How Wine Was Saved for the World* (2004). One of the reasons that Montalcino, originally famous for its white wines like ***Moscatello di Montalcino,*** switched to ***Sangiovese*** was the need to replant vineyards after the ravages of *la fillossera* and other diseases.

Fining agents
Fining agents are used to stabilize or clarify a wine. They work by various principles: electrical charge bond formation, absorption and adsorption. Once upon a time, animal products were used to clarify wines: ox-blood to soften **tannins**, egg-white to clarify, milk-based products (casein) and fish derivatives, crustacean shells, isinglass (fish gelatin from swim bladders,

a byproduct of the fishing industry) and so on. In 1987, all blood-related products were banned. The end of the 1990s and BSE (mad cow disease) marked the end of using beef products (gelatin, bone meal and so on). It was, in fact, the BSE crisis that pushed Europe into defining a common food safety protocol, and the foundation of the EFSA (European Food Safety Authority) in 2002.

In 2005, the use of all animal proteins was suspended while experts analyzed the allergenic aspects of protein residue from products used for clarification. It became important to make a distinction between additives and processing or fining agents. Although it was proved that filtration and other practices successfully rid wines of residues, all milk and egg derivatives were limited to 0.25 mg/l with an obligatory allergen declaration on the label from 2012. Isinglass has never been banned but is not generally used in **red** wines.

Finocchio

In July, wild fennel lines the roadsides in Montalcino, its bright-yellow flowers lacy and delicate. Fennel is a very Tuscan flavor. Just think of the delicious *Finocchiona IGP salami* that dates from the Middle Ages. Pepper was an expensive spice so the wily Tuscans replaced it with local fennel seeds in their cured pork preparations. Machiavelli was, apparently, a big fan. However, fennel is a complete no-no for wine pairing: up there on the naughty list along with lemons and artichokes. There is a verb in common use, *infinocchiare*, that means to swindle. It derives from the way that fifteenth-century tavern owners used to ply their customers with fennel before they gave them poor wines to taste. They also revived unsavory food with fennel galore, thereby tricking their guests. Here, wild fennel is used to give flavor to boiled chestnuts, ideally from ***Monte Amiata***.

Flavescenza Dorata

Flavescence dorée or golden flavescence... the name sounds so much more attractive than the phenomenon. Not unlike ***Xylella fastidiosa***, this disease is caused by an intracellular parasite transmitted by an insect vector. Symptoms include delayed or no **bud break**, patches on leaves and rubbery shoots. The vine deteriorates and dies. Though the disease is native to Europe, for a long time there was no spread until the right vector turned up. The dastardly *Scaphoideus titanus* probably hitched a ride to Europe from North America on rootstocks imported to tackle *fillossera*. This insect is making its way from the north of Italy to Tuscany and cases have been identified in Montalcino.

Monitoring and preventative treatments are key. National legislation was put in place in 2000, with obligatory controls in the vineyards, enforced quarantine for infected plants and penal consequences for producers who flout the rules. Recent studies have discovered that vibrational waves disturb reproduction between the leafhopper vectors that transmit FD, another example of **sexual confusion** at work.

Fogscapes
Three rivers – the Asso, **Orcia** and **Ombrone** – frame and delimit the **Brunello** territory, causing humidity that sometimes shrouds the top of the hill in fog, but more often veils the north-facing side. My father-in-law used to say that the people in Buonconvento *campano un anno in meno* (they live a year less) because so many of their mornings are spent in fog. Sometimes there is a basin of thick white fog below Sant'Angelo in Colle. You can usually make out the odd castle or *podere* floating without context, looking like boats on a great sea. In the winter, particularly, Montalcino often seems to be above the clouds and we all wish we were photographers.

Fortezza
Montalcino's impressive fortress is home to the oldest public *enoteca* in the world and visitors should make a point of walking the ramparts. Work began on the structure in 1361; it was designed by the Sienese who saw its potential as a military outpost. After the victorious defense of Montalcino from the Imperial and Medici militias in 1553, and the defeat and surrender

of Siena to the Medici in 1555, the *Fortezza* became the seat of the exiled republican government of Siena until 1559.

Siena honors this piece of history by giving Montalcino a place in the historical procession for the Palio. When Cosimo de' Medici I took over the city in 1559, Montalcino was the last independent municipality in Italy. Montalcino was annexed under the *Chateau-Cambrésis* treaty between France and Spain but it was not conquered in an armed attack. Draw your own conclusions.

From 1878 to 1896, Montalcino's dead were buried around the *Fortezza* walls, precisely in the buttress on the right that now hosts a small garden with benches and oak trees. In 1938, a huge restoration project began that included the exhumation of the old tombs.

Frost

In January 1985, Italy experienced tremendously low temperatures. In Venice, the lagoon froze over and in Florence, temperatures were recorded as low as -23° Celsius (-9.4° Fahrenheit). Around 90% of Tuscany's olive trees were damaged irreparably. In Montalcino, many farmers planted vineyards where the groves had been. Although the return on vineyards is slow, it is nevertheless quicker than the fifteen years it takes for an olive tree to produce. The **Brunello DOCG** (1980) along with the ***Rosso di Montalcino* DOC** (1983) were just beginning to take off. The landscape and destiny of Montalcino was altered forever by that icy winter. My own personal memory is of my father getting up from his chair in order to burn it in the fireplace.

April frosts in this area have been devastating in recent years (as low as -9°C / 16°F on 8 April 2021) and some producers have resorted to burning bales of hay, since the smoke can add a few precious degrees to the air temperature. Of course, the severity of the effects of a frost depends on what point the vine is at: eg if **bud break** has taken place, which in turn depends on the weather in the months preceding.

Fufluns
Fufluns was the Etruscan god of plant life, happiness, wine, health and growth in all things. His name derives from the Etruscan word for seed. He was ranked number nine of the sixteen ruling gods. Son of Semla and Tinia, he was worshipped in Populonia, one of the twelve city-states that were part of **Etruria**. The town was originally known as Fufluna, over time the name morphed into Pupluna, before settling as Populonia.

Funghi
Mushroom hunting is a bit like fishing in a forest ocean but with stationary prey and no bait. The weight of an empty basket is unbearable but little compares with the frisson and joy of recognition of coming across what you have been scanning the woods for, eyes roving from left to right.

There are over 250 species of mushrooms to be found in the Montalcino *bosco*, many of which are edible. The most sought after are porcini (*Boletus edulis*), known in English as ceps or penny buns. If heavy rains are followed by sun, mushrooms will come. The topsoil temperature should not go below 5° or 6° C. If you come across a cluster of Fiat Pandas on a woodland roadside, it means that the woods are pullulating with shroom hunters (*fungaioli*). There is an Italian saying '*anno fungato, anno tribulato*' which means that a good year for mushrooms is often not so great for agricultural products, including wine. The conditions that are ideal for mushrooms also provide the perfect environment for mold and mildew spores to thrive, including the dreaded ***peronospera***.

As far as picking is concerned, it will come as no surprise to learn that there is a series of rules and regulations. In Tuscany, there is a daily restriction of three kilograms per person, unless you happen to be resident in certain mountain ***comuni***, in which case you may 'harvest' as many as ten kilograms each day. If you forage outside your local area, you must invest in an annual permit. This sets back Tuscans in Tuscany all of €25. If you are not from Tuscany, you need a tourist permit which in 2024 had an annual cost of €100. There is also a size limit; porcini caps must be over four centimeters. Although exceptional, in 2023, a porcino that weighed in at more than a kilogram was found in the Montalcino woods.

A word about the downside of mushrooming; every year entire families are badly poisoned after eating mis-identified funghi. In Tuscany alone, halfway through the 2012 season, 168 people had been admitted to **hospital** for *intossicazioni*. Consider also that a common habit of *fungaioli* (mushroom hunters), excessively protective of their precious spots, kept secret for generations, is that they tell no one where they are going and even hide their cars from prying eyes. This behavior makes it hard to find them when they don't come home at sundown. In 2010, Italy-wide, forty-three mushroom hunters died in just fifty days. An old friend, in hospital for the last time, chose to spend our visit whispering his places into my husband's ear, since he had no children of his own for this precious inheritance.

Galestro

Galestro is not exactly a soil type but a shale-like sedimentary formation of clay schist. It is layered and friable, which means that it is great for drainage and root permeation, while retaining vital reserves of moisture. If you knock two pieces of *galestro* together, they splinter and break apart which, when it happens during natural erosion in the vineyard, helps aerate the soil. The fact that it flakes so easily means that it can release minerals and trace elements. **Sangiovese** made in soil with *galestro* generally shows depth and texture.

Gambelli, Giulio

Known as *Bicchierino* (or 'Little Glass'), Giulio Gambelli apparently hated this **nickname**, inherited from his great-grandfather in reference to their family-owned *trattoria* and the small glasses used there. Tuscany's undisputed and unassuming master of **Sangiovese** was born in 1925 and died in 2012.

He schooled so many producers in Montalcino and *Chianti Classico* that the adjective *gambelliano* is used to denote true *Sangiovese*. His career spanned almost seventy vintages and he is considered largely responsible for establishing the potential of monovarietal *Sangiovese*. After his death, an annual award was set up to recognize young **winemakers** who best embody Gambelli's ideals: exalting the typicity of a grape variety, the characteristics of the territory and the vintage.

Garbellotto

Considered by many as makers of the Rolls Royce of **barrels**, this family company are amongst the most established coopers of large barrels and have been established since 1775. They have state permission to fell oak trees in Slavonia, and own an enormous sawmill there.

Gatto, Alfonso
This writer and hermetic poet (1909-1976) wrote a beautiful panegyric to Montalcino in the 1962 issue of *Terra di Siena* (anno XVI, n.3) in which he commented on the town's 'endemic liberty' and, in passing, mentioned that it would never be a tourist destination.

Gelicidio
Haze ice is *gelicidio* in Italian, an evocative word that to the ear seems to collapse the **suffix** most commonly associated with murder and suicide with the word for ice. In March 2018, Montalcino suffered this phenomenon where **rainfall** forms instant and treacherous ice. Buds were encapsulated in ice bubbles and wires were coated with ice. Inhabitants of the town received megaphoned instructions to stay at home.

Geology
A great summary of Montalcino's geology can be found in Kerin **O'Keefe's** wonderful tome *Brunello di Montalcino: Understanding and Appreciating One of Italy's Greatest Wines* (2012). You should read it.

Brutally put, there is an enormous number of different soil types found here and the lack of a pedological continuum is well-documented. The producer Banfi has identified twenty-nine different soil types in its vineyards alone. The **whale in the vineyard** is a tangible reminder that this area was once under the sea. In Montalcino, in several prehistoric eras, the sea retreated and returned multiple times, churning up the soils with each tug of movement, scattering marine fossils all over the place like chocolate chips in cookie dough and folding deep soils into the accessible (to vines) top-layer. These effects of what is described as a geomorphological apocalypse is that in Montalcino it is perfectly possible for two vineyards on either side of the same road to have completely different soils, and the different tones of green within a single vineyard, seen from a distance, can indicate different soil texture, provenance and mineral content. Montalcino's specific and unique geology is the reason for the success of *Sangiovese* on this hill and of the necessary delimitations of the **DOCG**.

Gian Gastone

This is the name given affectionately by the *ilcinesi* to the 1564 statue which currently stands under the covered loggia below the clocktower. The sixteenth-century sculpture is attributed to Giovanni Berti, who was known to have worked in Rome restoring ancient sculpture. His only other known work is a 1570 marble relief in the *Palazzo Sansedoni* in Siena. This Michelangelo-esque cameo has as an inscription 'IO BERTIUS ILCINENSIS SCALPEBAT' – a clear statement that Berti considered himself a Montalcinese rather than a Tuscan.

The statue was originally in the *Cappellone* and celebrates Cosimo I (1519-1574), who was the first Grand Duke of Tuscany, from 1569 until his death. He was the great-great-grandson of Lorenzo the Magnificent and, amongst other things, commissioned the building of the Uffizi in Florence. He also defeated the Sienese after a fifteen-month siege in 1555 and annexed Montalcino in 1559, effectively freeing it from dominion by Siena.

Cosimo I brought wellbeing to Montalcino, and fostered the rebirth of the town after the *assedio* by allowing the town to maintain its idiosyncratic tax permits. The sculpture was commissioned to celebrate a visit of Cosimo to Montalcino that never happened. On the day of its inauguration, there were choirs of singing children. You may be familiar with the more renowned statues of Cosimo I in Florence and Pisa, commissioned after his death by his son, Grand Duke Ferdinand I. The former, by Giambologna in 1594, is in *Piazza della Signoria* and shows Cosimo I on horseback, the first equestrian statue of a ruler. The other, in *Piazza dei Cavallieri* in Pisa, shows Cosimo I standing proudly and dates from 1596.

Considering all this, two questions come to mind. Firstly, why was this statue moved to such an uncelebratory location? It seems likely this took place after the **Unification of Italy** in 1861, but it does not strike me as a position of honor, tucked away, out of sight, with children bouncing balls off his extremities summer after summer. More interestingly: why is the statue known as *Gian Gastone*, when this is the first name of the seventh and last Grand Duke of Tuscany (1671-1737)? *Gian Gastone* is vilified by contemporary anecdotes as a depraved and debauched dipsomaniac with innumerable unpleasant habits that included regularly vomiting on his dinner table and refusing to have his bedsheets changed. The town **nickname** is so widepread that the internet labelling of images of this statue often wrongly ascribe it as depicting *Gian Gastone*, in spite of it having been chiseled into existence over a hundred years before he came into the world. This joke is so long-standing that hardly anyone is aware of its implications and it can be interpreted variously as a sign of Montalcino's 'endemic liberty', civic pride and complicated relationships with both Florence and Siena – as well as a love of nicknames.

Giardini dell'Impero

These are the gardens below the ***Fortezza***; the benches are a sublime spot for sunset watchers.

Gin

In 2021, two friends, Simone Meattini and Gianluca Turchi, created Montalcino's own London dry gin, *Ilginus*. The botanicals are juniper, olive leaves, **cypress** berries and wild asparagus and fennel (*finocchio*), all typical and evocative tastes of this territory.

Ginestra

This beautiful shrub, *Ginestra odorosa*, has bright yellow flowers that smell of honey and Sicilian white wines. The plant is related to Scottish gorse and to the many species and subspecies of broom. In spring there are rowdy splashes of yellow wherever you look on the slopes of Montalcino, and the aroma is intoxicating.

Giro d'Italia
This epic annual **bicycle** race began in 1909, six years after the *Tour de France*. Montalcino hosted a *tappa* (stage) in 1987, 2010 and 2021, when the finishing line was in the center of the town. After zipping through the *Val d'Orcia* and its postcard views, there were 35km on **strade bianche** (gravel roads). Anyone who has ever driven the rough roads from Castiglion del Bosco to Montalcino, got lost below **Camigliano** or on the treacherous Sesta, will be able to imagine what this entailed: high puncture risk and exciting viewing. The official Giro site describes the Montalcino stage as 'as challenging as tackling the Alps.' In 2010, this section was the most watched stage of the Giro when Cadel Evans sprinted to victory through the rain and all the cyclists were slick with red mud from head to foot.

Glassware
Yes, it does make a difference, and yes, if you are serious about your *vino* then you ought to invest in some good glasses. Some people swear by Zalto or Riedel. In 2021, the **Consorzio del Brunello** created a glass specifically for **Brunello** called Senses T-made 70 Excellence. Make sure you do your research sitting down; the cost may make you swoon.

Gorelli, Gabriele
Gabriele Gorelli is from Montalcino and in 2021 became Italy's first Master of Wine. There are currently 414 Masters of Wine, living in thirty-one different countries (502 people have passed the MW exam since 1953). Born in 1984, he has deep roots here and has his own town **nickname**. He features as a cameo in *Vanilla Beans and Brodo* during the Etruscan dig at **Poggio Civitella** and was one of the last children born in Montalcino **hospital**, receiving the elusive **F402** fiscal code.

Grandine
It is possible to stipulate an **insurance** policy for *grandine* (hail), though few producers can even enunciate the word without having to clutch the

77

front part of their trousers. Hail, along with **heat spikes** and late **frosts,** has become a real issue, and it is very sobering to lose nine months' work in a five minutes' battering. There was terrible hail in August 2023.

Granfondo del Brunello

This was a national mountain bike competition held in Montalcino from 1990 to 2020. It has now been replaced by ***Val d'Orcia Gravel***, which takes place in the third week of October. In the early days, the incentive to signing up early was a bottle of ***Rosso di Montalcino.***

Grappa di Brunello di Montalcino

Some people are rightly wary of *grappa*, traumatized by past experience. And it is certainly true that some *grappas* might be best used to remove nail polish or start fires. However, a good *grappa* is a thing of beauty. It is made from distilled **marcs**, a byproduct of winemaking. It follows that when a *grappa* derives from a **denomination** wine, all the care that was taken of the fruit and vinification processes is reflected in the primary ingredient for the *grappa*. Add a great distiller and some wood aging (in a wine-impregnated cask) and you may be pleasantly surprised. The first monovarietal *grappa* in Italy were made in the 1970s by Montalcino producers. A *cafe corretto*, a quick pour of *grappa* into an espresso coffee, is invigorating on a winter morning, but the best time for a small glass of *grappa* is as a *digestivo* after a long Sunday lunch. Set aside your prejudices and give it a go.

Grecale

This is a north-easterly cool **wind** that can come in gusts, and is most typical in late autumn and winter.

Growth System

A growth system is the method used to train and trellis the vines with the assistance of wires and posts. Vineyard trellises are an important part of the vineyard design. As far as **Brunello** is concerned, this is not regulated by law, yet *Guyot* and *Cordone Speronato* prevail as being the two best for *Sangiovese*. The young vines are manually bent and tied at a determined height and then pruned accordingly, based on an ideal distribution of clusters.

As Montalcino experiences hotter summers, many producers are switching from *Cordone Speronato* to *Guyot*. The choice of growth system determines the **canopy management** of the years ahead and can affect yield. The ideal growth system will allow for balance between foliage and photosynthesis and fruit production.

Guyot

This **growth system**, once extremely common in Tuscany and, as part of a Tuscan renaissance is now being reintroduced in many vineyards in Montalcino. It was developed by Jules Guyot in the second half of the nineteenth century, and is particularly suited to clay or stony soil because it allows the vine to make the most of whatever resources are available. It involves more annual maintenance and is less intuitive to prune well but has the advantage of regularizing production and extending vine-life; it helps prevent disease and can offer protection from **frost** damage since it delays budding.

Heatwaves and Heat Spikes

Of the two, a wave sounds more pleasant to endure, though I doubt the grapes pay attention to these linguistic subtleties. A heat spike or heat burst is a localized phenomenon that follows a thunderstorm and usually takes place at night. There is a rapid increase in surface temperature, dramatic decrease in surface dew point, and on some occasions, damaging **winds**. This was the case in August 2011 in Montalcino.

Heatwaves, by contrast, are sustained periods of excessive heat and they have become part of summer in recent years. There were three in 2023, with temperatures in ventilated Montalcino hitting the forties (104°F). The first was named *Nero* (after the Roman emperor accused of starting the Great Fire of 64CE in Rome); the others *Cerberus* and *Caronte*, respectively the three-headed dog guarding the gates of hell and *Charon*, the boatman of Hades. Tropical cyclone names are assigned according to an internationally agreed rota, but heatwaves have no official naming system and meteorologists bicker about whether they ever should. These mythological names, that dominate conversation and front pages, have been given by an Italian weather website to describe extreme heat since 2017. They have been reprimanded for being too sensational, which is, of course, why the public and newspapers alike adopt them with enthusiasm.

Hectoliter, Hectare and Hectogram

All these measurements have their root in the Greek *hekaton*, meaning one hundred. In Italian, the h and the c have been lost by the wayside to partially reappear in the abbreviations, for maximum confusion. *Ettolitro, ettaro* and *etto* become HL, HA and HG.

A hectoliter is one hundred liters, and is the measurement for storing, buying and selling wine. Grapes are harvested in **quintals**, just to keep things interesting.

A hectare is 100m2 and is bigger than a football pitch. There are one hundred hectares in one square kilometer. An acre is about 0.405 hectare and one hectare contains about 2.47 acres. Vineyards are measured in hectares. In Montalcino, there are 258 wineries that have less than ten hectares of vineyards – the vast majority of producers are in this bracket.

A hectogram is one hundred grams or about 3.5274 ounces. The word *etto* (plural *etti*) will often be heard pronounced by indomitable grannies in conjunction with orders for sliced meats and cheese.

Hemingway, Ernest
Lived to write, thanks to the courage of a young *ilcinese*, Fedele **Temperini**.

Hospital
Montalcino's hospital, *Santa Maria della Croce*, has been serving the community for nine centuries, during which time it has changed both location and function. In 1964 it had 105 beds and thirteen different departments, including cardiology and orthopedics. It is currently housed in a beautiful structure but no longer has a maternity ward, amongst other losses. It has only eleven beds overall, and severely reduced services.

There are plans afoot to return it to its former glory as a community hospital which would be a most welcome development. At present, the nearest hospitals are Siena or Montepulciano, at least forty minutes away depending on traffic. If you find the drive tiresome on holiday, be grateful that you are not in active labor, stuck behind an old man with a hat in a Fiat Panda. These distances mean that rural ambulances are more equipped than urban ambulances, that many town squares have a defibrillator, and that emergency helicopter rescue and transfer is not uncommon. The helicopter service is called *Pegaso* after the mythical winged horse. In Montalcino, helicopters land on the football pitch in front of the **Fortezza**, or in the several other heliports located on key areas of the hill.

ICQRF

Pronounced Itchy-coo-erre-effe, this acronym identifies the Agricultural Fraud Repression Squad, established in 1986. It is the Law Enforcement Body of the MASAF (Italian Ministry of Agriculture, Food Sovereignty and Forests), with a remit to prevent any breach of European and Italian laws relating to food production. It controls, inspects, administers fines and works closely with police and prosecutors. Every winery is assigned an alphanumeric ICQRF number which can be used to identify where the final product was bottled. Until 2020, it was obligatory to have this number on the cork and capsule of every **bottle**, but this no longer applies.

Idonietà

This certification is the last step required by the **DOCG** regulations before the wine can stop being **AAD** and start being itself. Gamblers can request the *idonietà* after bottling, but it is considered more prudent to have the inspection before the wine is bottled. ***Valoritalia*** employees come to syphon off four **bottles'** worth of wine in anonymous bottles that they cork themselves with anonymous corks. Two bottles remain with the property and two are taken away, one for lab analysis and one for a tasting panel. Producers can request the long list of potential tasters but will never know who is tasting their wine. Similarly, the tasters do not know whose wine they are tasting.

I Giorni della Merla

The last three days of January are known throughout Italy as *i giorni della merla* – the days of the blackbird. According to tradition, these days are always bitterly cold and used to be the coldest three days of the year. Legend has it that the blackbird was once brilliant white but its plumage turned black forever after three days of sheltering in a chimney to escape the perishing cold. Temperatures on these three days are supposed to give an indication of the spring to come. If it is warm, spring will be late, whereas if it is cold, there will be a *bella primavera*. The **Candelora** is another indicator if winter is over or not, though global warming has interfered with these once-reliable touchstones.

IGT

Indicazione Geografica Tipica is a relatively loose wine **denomination**, created in 1992. It exists throughout Italy, followed by a specification of a region or a province, eg *IGT Toscana, IGT Maremma Toscana*. Essentially, it means that the grapes have been sourced from the mentioned area. Restrictions about alcohol and **yields** are not stringent. A free-for-all in other words. There are over 120 IGT in Italy.

Il Bosco della Ragnaia

This is a woodland park and garden created by American artist Sheppard Craige in **San Giovanni d'Asso**. Work started in 1996 and continues today. It is divided between a shady woodland part and a more structured valley area below. There are enigmatic and whimsical installations, inscriptions and carvings, all loosely based around the themes of uncertainty, nature and time.

Entry is free and the park is open from dawn to dusk... when it is open. Check before making the drive.

Ilcinesi

The official demonym for the inhabitants of Montalcino is *ilcinesi* (singular *ilcinese*). In Italian, this sounds comically similar to *i cinesi*: the Chinese.

In 2017, identifying the meaning of *ilcinesi* was one of the challenges posed in a national quiz show, *L'Eredità*. Unfortunately, the contestant fell into the obvious trap.

Il Prato

In what might seem a consummate Montalcino piece of humor, the gardens at the bottom end of town are known as *Il Prato* (the meadow), as is the bar there. In summer, people rest on the granite benches under the shade of the lime trees and watch visitors trying to work out the vagaries of the bus timetable (when is a *feriale festivo*? Does the *scolastico* run on Saturdays?). The *giardini* are opposite the *Bar Prato* where bus tickets can sometimes be purchased but on no account is change given for the parking meters all around. The *giardini* contain outdoor tables for two establishments and the waiting staff have a terrible time of it, carrying plates across the road while dodging the traffic. There is a small fish pond, much loved by Montalcino's children, and very well-kept flowerbeds divided by paths. But even with all this, on no account could it be described as a meadow.

Imbottigliamento

Article 5.12 of the **DOCG** regulations specifies that vinification, wood-aging, bottling and **bottle**-aging must all take place within Montalcino territory. Bottling ***Brunello*** can be dictated by prosaic concerns such as freeing a barrel or meeting sales commitments. Many producers choose to bottle during the summer months to reduce potential oxidation of the wines since oxygen dissolves in wine at lower temperatures. It is vital to bottle as far away as possible from harvest and fermentation when the cellar is full of stray **yeasts** and potential sources of future problems. As per DOCG regulations, *Brunello* must spend a minimum of four months bottle-aging before release in January while ***Riserva*** must spend six months. Bottling in summer means that the wines are ready to be sent out when they can be commercially released, on the fifth January after harvest as per Article 5.17.

Insurance

Fortunately, insurance is a possibility. EU agricultural policy reimburses a significant part of the cost of the insurance premiums but the actual indemnity is quite low, compared to the value of grapes and lost harvests in

Montalcino. Sadly, the list of insured weather events has been lengthening over the years. At present, in the catastrophic calamities, you find flooding, **drought** and **frost**. **Hail, wind,** excessive **snow** and rain are listed in the frequent section. In the so-called ancillary adversities, that read like a grandmother's finger-wagging admonition to a poorly dressed *nipotino*, there are all the issues connected with higher temperatures: *colpo di sole* and *sbalzo termico*, along with **heatwaves and heat spikes.** At present, there is no provision for the damage wreaked by wild boar and deer.

Invaiatura

Termed **veraison** in English (straight from the French, *naturellement*), this is the moment in which the grapes begin to change color and consistency and is a visual cue that harvest is close. This is a perfect time to use purple as a verb. Each grape matures at its own pace. A bunch, blushing grape by single grape, is a sight to behold; a cluster of **Sangiovese** can look like a bridal bouquet. As the color changes, the berries are softening from hard, green, acidic bullets into grapes. Chlorophyll is replaced by anthocyanins. Organic **acidity** levels decrease and sugar levels rise.

After *invaiatura*, berries swell up in size as sugars accumulate, while flavors and aromatic compounds develop.

Inventory Loan

The *Monte dei Paschi* bank was the first to normalize the practice of inventory loans in Montalcino. Wineries can leverage credit based on an external evaluation of the **AAD Brunello** in their **barrels** at the winery. It is a popular way for wineries to monetize their liquid stocks, since the credit reimbursement only begins when the *Brunello* is released, on the fifth year after harvest. In central Italy, future productions of *prosciutto* and *parmigiano* can also be financed in this manner.

The Italian word for this kind of loan is *pegno* and the practice was established in Roman times. Pawning is still regulated by banks in Italy and most financial institutions have a *Monte dei Pegni*. The first to open its doors was in Perugia, in 1462. They have been doing a roaring trade in recent years.

Irrigation

A newly planted vineyard can be irrigated until the vines are producing grapes. Once the *barbatelle* are grown, they are on their own, and a bit of hardship is a good thing, forcing the roots to go deeper to access water.

After the scorching 2003 vintage and a decade of lobbying, the Italian Ministry for Agriculture published a circular in April 2013, clarifying that *irrigazione di soccorso* is automatically permitted for **DOCG, DOC** and **IGT** wines when not mentioned in their regulations. Until this point, dry farming had been obligatory. This specification had become necessary due to the combination of increased temperatures and decreased **rainfall**. The ***Brunello*** **DOCG** ***disciplinare*** now specifies that any vineyard practice *di forzatura* is strictly forbidden but that *irrigazione di soccorso* is permitted.

Emergency irrigation is, by definition, an extreme measure that, when used correctly, has the purpose of safeguarding the harvest or the vine itself. There is no risk of it being used to increase the yield of grapes (thus impacting quality) since **denomination** wines have legislation in place that limits **yields** (to guarantee quality). In principle, emergency irrigation seems like a simple solution but it definitely has to be a rare intervention and only in vintages characterized by extreme and prolonged **drought**. Watering when the vines are already in difficulty is counterproductive (think of the catastrophic effects of giving a dehydrated person too much to drink too quickly) and, in 'normal' vintages, some level of ***stress idrico*** is acceptable and can even be positive for the vines, for example in driving the roots deeper.

This is a sophisticated distinction that is also connected with winery size and budget. The business of emergency irrigation can be seen as being undemocratic when only the larger wineries have existing lakes or the funds and available lands to create them. The logistics of procuring 300-400 cubic meters of water per **hectare** are no small feat.

J

The Italian alphabet officially has twenty-one letters and J is not one of them, as my son James knows every time he has to spell his name. Depending on the age of the interlocutor, he either has to mention a football team or describe his initial as '*i lunga*' – long i. The other 'missing' foreign consonants are K, W, X and Y which generally appear in loan-words, Latin words, proper names and a very few regional words.

In Italian, the letter G makes a J sound, as in *Luigi* and *gelato*. The J itself occasionally makes an appearance in nineteenth and early twentieth century Italian on those occasions in which, before a vowel, an 'i' behaves like a consonant. Clever clogs may already be thinking of one **Brunello** producer with a J in the historical spelling of the property name. Clue: *La Dolce Vita*.

Kermesse

This word, borrowed from sixteenth-century Holland to describe a religious festival, derives from the words *kerk* (church) and *mis* (mass). It is invariably used by Italian journalists in conjunction with **Benvenuto Brunello** each November, in its acquired sense of a spectacular event featuring many different acts. It is the only time of the year I ever hear this word.

Large Formats
Article 8 of the *disciplinare* lists the accepted volumes of **bottles** for *Brunello*: 0.375 litre (a half bottle); 0.5lt (two-thirds of a bottle); 0.75lt (standard bottle); 1.5lt (Magnum); 3lt (Double Magnum); 5lt (Bordeaux Jeroboam); 6lt (Methuselah, the equivalent of 8 bottles); 9lt (Salmanazar, 12 bottles); 12lt (Balthazar, 16 bottles); 15lt (Nebuchadnezzar, 20 bottles) and 18lt (Melchior, equivalent to 24 bottles or two cases). There are also bigger formats, going up to the magnificent Melchizedek, which contains 30 litres or 40 bottles, as well as the neat 570ml Winston Churchill (the ideal measure of *Champagne* for morning quaffing), but these are not permitted for *Brunello*.

It is worth overriding any cellaring issues resulting from their awkward dimensions since, as bottle-size increases, so does longevity. Ageability for some *Brunello* vintages is a wonderful thing. Add to that the frisson of rarity and the fact that many of these larger formats have to be hand-labeled or even hand-bottled, and you will see why some people go weak at the knees for bigger bottles.

La Madonna
Witness Montalcino synecdoche at work here. *La Madonna* refers not just to the church on the outer **city** walls, dedicated to Montalcino's patron saint, but to any part of the walk along those walls. It is quite common to give one's location as *La Madonna*.

La Titina
If you have been to *Il Prato*, you will probably also have encountered Mario Marconi, known in Montalcino as *La Titina*. This **nickname** was given to

him over seventy years ago when he was a fifteen-year-old builder's assistant, working on the *Fortezza* walls for the almighty sum of five hundred lire a day. As often with Montalcino nicknames, the *soprannome* came from a chance exchange. *La Titina* was a 1917 song made famous by Charlie Chaplin in *Modern Times*. The head builder, a man so prodigious in weight he needed two boards to his bench, was apt to sing it as he worked. All it took was Ugone stopping mid-song and calling young Mario '*La Titina*' once and now, ever since, Mario has been known by that name.

He is a dapper man with a moustache, who wears a panama and braces in summer and a tie in winter. He keeps the gardens in order and at Christmas sets up the town's *presepe*, an enormous nativity scene that spreads across the gardens.

La Torre dei Pomodori

If you have driven from Siena to Montalcino, you will have come across this massive structure, created for the industrial freeze-drying of fruit and vegetables. An inevitable experience for all visitors to this area is a glimpse of the Tomato Tower, also known as the less poetical *ecomostro*. The former IDIT plant (Isola Tressa Dehydration Industry) stands seventy meters high, dominating the **Val d'Orcia** and visible for miles. It was built in a record two years and was officially opened in 1961 with coverage in the national media. Governmental grandees and the Archbishop cut the ribbon in a moment of great excitement and absolute confidence in the economic recovery of the area.

Tragically, the tower was only operational for two years. Due to a design defect, it was deemed unviable and production ceased completely in 1966. Privately owned, it still stands, slowly falling into decrepitude. There are often rumors of plans either to renovate or demolish it.

Leccio

This is the local word for the holm or holly oak, *quercus ilex*, an essential part of the Mediterranean *macchia* and one of the three trees most suited for truffle orchards. The name Montalcino derives from **Mons Ilex**, the mountain of *lecci*, and the *bosco*, once essential to local economy, is still part of the landscape and covers huge areas of our views. It is an evergreen which means, of course, that the hills are never bare.

Leccio d'Oro
This is an award given by the **Consorzio del Brunello** to those who have furthered the name and fame of **Brunello** in the world beyond Montalcino.

Libeccio
The *Libeccio* **wind**, its name deriving from the Greek for 'from Libya', comes from the south west and brings warm, wet air from the sea. It is a summer wind and in Montalcino often predicts rain. First use: 1667.

Liber
Liber Pater was the name that the Romans often applied to the Greek god **Dionysus**, also known as **Bacchus**. Liber and his wife, Libera, were agriculture-fertility deities in pre-Roman Italy, with particular connections to viticulture.

Lightweighting
Apparently, this is now a verb rather than just an insult. Lightweighting refers to the practice of choosing lighter **bottles** in order to reduce carbon footprint in transport.

Macchia

Macchia refers to the Mediterranean bush, composed of dense evergreen shrubs and small trees. To confuse your interlocutors, throw the word *maquis* into conversation. In Montalcino, this wooded area is composed of many varieties, including holm oaks, **albatro,** myrtle, and juniper.

Maestrale

The *Maestrale* **wind** comes in from the north west. Unlike the mighty mistral in France, by the time it reaches Montalcino, the *Maestrale* is neither strong nor feared.

Maggese

This Renaissance term for ploughing a field and leaving it fallow as part of crop rotation has come to mean any time of regenerative rest. After uprooting a vineyard, a period of one to three years of *maggese* is necessary before replanting. The practice, and its periodic necessity, extends beyond viticulture to other forms of agriculture and to humankind.

Malolactic Fermentation

This is the secondary fermentation (now known as malolactic conversion) when the sharp, tart L- malic acid turns into softer L+ lactic acid, releasing carbon dioxide in the process. There is a change in the mouth-feel of the wine and an increase in **pH**. Some wineries allow this to happen naturally in spring when temperatures begin to rise; others prefer it to take place as soon as possible and may even inoculate to this end. Wines prior to malolactic conversion are less microbiologically stable and cannot be stabilized by adding sulphur dioxide since this impedes the process. Abbreviated to MLF.

Manine della Vigna

As the vines grow upwards, they anchor themselves to the training wires with their tendrils. The agronomical term is *viticci* but in Montalcino they are known as *manine* – little hands – since they look as if they are clinging to the wires for support.

Manodopera

Who is it doing the all-important work of tending to the vines: the ***potatura***, the ***stralciatura*** and so on? *Manodopera* – workforce or manpower – is the term used for manual labor of all kinds. Due to complicated labor laws and stringent health and safety requirements, it has become increasingly common to rely on third-party labor teams. Many producers in Montalcino rely on their full-time staff for significant operations such as pruning and shoot selection and call in the teams for harvest and other less-specialized interventions. These *squadre* are generally made up of people from Eastern Europe, North or Sub-Saharan Africa; some are resident year-round in Italy and others come with a work permit for the months of the ***vendemmia***.

Marcs

Marcs or pomace, in Italian, ***vinaccia*** are a byproduct of winemaking and are used to make ***grappa***.

Marinetti, Filippo Tommaso

At the second *Mostra Mercato Dei Vini Tipici d'Italia*, held in Siena in August 1935, the futurist poet Filippo Tommaso Marinetti (1876-1944) leaped onto the table during the gala dinner to celebrate a poetry competition, and, raising his glass, shouted '***Brunello è Benzina!***' (Brunello is petrol!). The sentence was in full harmony with the Futurist movement and became the motto of the Fair.

Marinetti is also accredited by some for the first *Insalata Caprese*, served as part of a futurist dinner in the 1920s – certainly a better legacy than his support of fascism.

Massa
Nothing to do with Formula One, this word refers to a partial or complete blend of one or more **barrels** of the same vintage to create a final wine. See also **Coacervo**.

Mattonella
Every year, since 1992, the **Consorzio del Brunello** has commissioned an artist or public personage to create a tile or *mattonella* to commemorate the latest harvest. Each tile is cemented into the wall of the town hall with maximum suspense and pomp. Look out for 2009, created by the forces behind the best-selling *Drops of God*, a Japanese manga series about wine, described by *Decanter* in 2009 as 'arguably the most influential wine publication for the past twenty years,' Chef Cracco's inscrutable egg (2021), or cashmere king Brunello Cucinelli's **Dionysus** (2022). You will find the complete list at the end of this book.

Maturazione
There are two kinds of ripening: technological and physiological. The former refers to the relationship between sugar content and acids, whereas the latter indicates the ripening of the grape components, ie skins, seeds, **tannins** and aromas. In a perfect vintage, the two kinds of ripening coincide perfectly. Unfortunately, climate change is driving a wedge between the two and grapes often reach technological maturation before they are physiologically ripe.

Mezzadria
This ancient system of feudal land tenure, not unlike share-cropping, dates from the ninth century. A farmer paid a proportion (originally half) of the farm's produce as rent and lived on the property, with the entire family working the land. The *case coloniche* that we associate with Tuscany derive in part from this system that permitted aristocratic city families to invest in agriculture remotely. In Tuscany, the 1785 reforms by Grand Duke Pietro

Leopoldo consolidated *mezzadria* as the most common form of land management. The reforms included incentives to build new farmhouses to improve the living conditions of the *mezzadri*. In the nineteenth century in Montalcino, 90% of farmland was *mezzadria*. After the Second World War, the Italian Communist Party sought to improve social and economic conditions in the countryside and the system began to collapse. In September 1964, new *mezzadria* contracts were forbidden but those already stipulated were allowed to continue unmodified, though with an improved division of forty-two percent to the land owner and fifty-eight percent to the '*mezzadro*'. In 1982, a new law had all *mezzadria* contracts morph into agricultural contracts.

In Buonconvento, there is a museum dedicated to *Mezzadria* and it is significant to Montalcino insofar as some of the great indigenous producers have their roots in this system. Stefano Cinelli Colombini, from the historic *Fattoria dei Barbi* estate, considers the end of *mezzadria* as a temporary nail in the coffin for Montalcino's growth (along with the building of the *Autostrada del Sole* which effectively cut Montalcino off from transit).

Microclimate

Anyone who lives in Montalcino has daily proof of the variety of microclimates on these slopes. It hails in terrifying stripes or rains in concentric bands around the hill so that you are constantly switching the wipers on and off as you drive from A to B.

The best time to verify the different microclimates is in spring, when the first almond trees are in bloom, splashes of pink standing out in the military shades of winter, here and there from Sant'Angelo in Scalo up to Sant'Angelo in Colle. Any higher, and the trees are yet to blossom. Next, the *mandorle* flowers will be joined by bright yellow mimosa, blossoming in the most protected spots first. There can be a month's difference between the start dates of harvest for low-lying and high-**altitude** estates, and 5°C temperature differences between the Valdicava area and Sant'Angelo on the same day at the same time.

Miele

Montalcino is a **city** of honey, amongst all other things, and has been recognized for its production since the 1970s. *Acacia, Sulla, Millefiori* and *Castagno* are the typical varieties. If you haven't sampled chestnut honey with local *pecorino* this should be remedied as soon as possible.

Millerandage

To be said once a day in the month of September, ideally in the context of 'as you can see, no millerandage is present.' Sounding like a cross between a Victorian heroine and a kinky bedroom activity, this fabulous noun refers to uneven size and ripeness of individual grapes within a single cluster. *Sangiovese* is particularly prone to uneven ripening. This can be caused by nutritional deficiencies in the soil or by humidity and rain during flowering.

Drought during the period between fruit set and **veraison** can also result in this phenomenon. The outcome is wines that taste unbalanced or 'green'. I discovered recently that millerandage is also delightfully described as 'Hen and chicken' and 'Pumpkin and peas' and, more prosaically, as 'shot' or 'shattered'.

MOG

Stands for Material Other than Grapes and refers to the debris that can be inadvertently harvested along with the grapes. This is a good acronym to have at hand when at the sorting table and a stink bug appears amongst the grapes.

Moglio, Giovanni

Known as the Sienese Savonarola, Giovanni Moglio (1500-1553), was a native of Montalcino. He died an excessive death in Rome, hung and burned at the stake as a heretic, after refusing to acknowledge the hegemony of the Catholic Church. There is still a street commemorating his name in Montalcino.

Mons Ilex

Mons Ilex means 'the hill of the holm oaks' in Latin and is considered by many to be the root of the word Montalcino. Holm oaks (also known as holly oaks) are large evergreen trees and are very common in Tuscany, as in much of the Mediterranean. Latin scholars will know that *ilex* is similarly ubiquitous in the *Aeneid*; Dido and Aeneas adorn each other with *ilex* garlands, and the golden bough for entrance to the underworld is on an *ilex*. In Italian, an oak tree is a *quercia*, but a holm oak is a **leccio**.

The other possibility is that the word Montalcino derives from *Mons Lucinus*. *Lucus* in Latin means sacred or small wood. This is how Montalcino was described in the first document that refers to the town, dating from 29 December 814. Either way, the town crest, embossed on lamp-posts and manholes, shows a solitary oak on a symbolic mountain. If you have come across a sculpture of a strange iron tree on a heap of rubble in the inner courtyard of the town council, now all will be clear.

Monte Amiata

This extinct volcano, 1738 meters above sea level, dominates the **Val d'Orcia** and the Maremma and separates the provinces of Siena and Grosseto. It is a ski-resort, fiery with chestnut woods in autumn. It also has a relationship with half of Montalcino's producers, described variously as a shield, a weather-breaker and protector from hail and storms.

Months

Perhaps it will come as a surprise to learn that, for certain types of employment contract, the Italian salary year is comprised of fourteen months rather than twelve. There is a double payment in July (for August holidays) and in November (for Christmas). These payslips are known as The Thirteenth and The Fourteenth and are much anticipated. This was one cultural difference that I found very easy to adopt when I first started working in Montalcino.

Mosaic Roundabout

Some might think that placing an artwork at a five-point intersection might be a distraction, but they were not consulted on this matter. On arriving at Montalcino, there is a Bacchic figure made of mosaic on a dolmen. In fact, it is entitled *Dioniso*, Italian for **Dionysus**. Most visitors are probably too busy trying to choose the right road to appreciate the artwork fully, but it might be worth a second loop round. The design was donated by artist and producer Sandro Chia.

Moscadello di Montalcino DOC

Until the end of the nineteenth century, this was the wine for which Montalcino was famous, with numerous references from the sixteenth century onwards. Thanks to its unmistakable sweetness, *Moscadello* was much loved, but its rise was 'interrupted' due to vineyard diseases (including *fillossera*), eventually resulting in the demise of the *'moscadellaie.'* *Moscadello di Montalcino* became a **DOC** in 1984. There are ten or so producers currently making this wine for a total of around fifty thousand **bottles** in three different expressions: sparkling, classic and late harvest.

There is an alternative universe in which Montalcino is still known for its white wines. Ezio Rivella (1933-2024), Montalcino figure and briefly President of the **Consorzio del Brunello** (2010-2012), was on the record in a town meeting suggesting that producers uprooted *Sangiovese* in order to supply grapes to Banfi for the production of a fizzy *Moscadello*, *Il Ricciardello*. At one point, Banfi had 350 **hectares** of *Moscadello*. This project was abandoned in 1999 and they now have under seven hectares.

Museo della Comunità di Montalcino e del Brunello

The museum is open weekday mornings from March to October and by request at other times, by contacting *Fattoria dei Barbi*. It is over 1,500 m^2, containing numerous exhibits and testaments to life in Montalcino, both from the pre-***Brunello*** times characterized by ***mezzadria*** and from the origins of *Brunello* in its pioneering aspects.

Must

The word comes from the Latin *vinum mustum* which means young wine. In Italian, *mosto*, this is the freshly pressed grape juice that still contains the skins and seed. It is the very first step for grapes becoming wine. Tasting *mosto* requires a whole different palate training and, although deceptively like a fresh juice, it has powerful laxative effects.

During this first phase of fermentation, CO_2 is released: an unforgettable, sickly smell. In 2010, we lived above the cellar and had a harvest baby; that unforgiving aroma and the joy at the safe arrival of our third child are forever mingled.

Nacciarello

In rural geography, it's common for each curve in the road to have a name, making it as easy to locate for locals as a street number is for New Yorkers. *Nacciarello* is the name given to the area that leads from Montalcino to Castiglion del Bosco, close to the crossroads for Castelgiocondo.

Here you will find a well-tended memorial *cippo* (a broken column symbolizing interrupted lives) placed there for six Montalcino men (one just a boy) who were killed by a German mine in the Second World War. The surnames are familiar to anyone living in Montalcino.

Napa

A city in California, in the eponymous county and wine region, twinned with Montalcino since 2021.

Neckstrips

The **DOCG** neckstrips, or *fascette*, once pink, are now taupe. They are vintage-specific and estate-specific and correspond to the exact number of **bottles** that have been certified as *Brunello*. Producers pay for them and must keep track of the numeration.

At the end of December 2009, the **Consorzio del Brunello** launched a new traceability procedure via text message. This has since turned into a consultation via their website. By inserting the numbers on the neckstrip, final consumers can check the wine they are drinking is certified *Brunello* and has been produced by the estate on the label (and is the vintage as labelled). This is part of an Italy-wide move towards transparent practice and to make wine fraud more difficult. A wealth of other technical details is available, including **acidity** and **ABV**.

Nicknames

Often generational, mostly male, entire books have been written on this subject. The *soprannome* or nickname is integral to Italian life. The Italian word literally translates as 'overname'. Many *soprannomi* appear harsh or cruel but often are experienced as a term of endearment and an expression of belonging. In Southern Italy, where family naming traditions are strict (eg first child named for grandparents and so on), the use of nicknames can be explained as being a way to differentiate amongst extended families where many members share the same given name. A glance at a Montalcino cemetery will show that non-religious and particular names were rife last century, in keeping with the left-leaning, anti-clerical history of this region. Tuscany's history of communism and individualism meant a generalized rejection of Catholic and Roman names until relatively recently. And when I say particular, I mean that a baby-namer would have a field day. Within my husband's family, we have Azzorati, Iader and Eufemio.

Returning to nicknames, let me remind you of the Renaissance painters, identified by birthplace, like Parmigianino, Perugino or Caravaggio, or by physical appearance: Rosso Fiorentino with his fiery hair; Pinturicchio, the diminutive painter responsible for the wonderful Piccolomini Library in Siena; or Botticelli. And let us not forget Sodoma.

Inhabitants of all Montalcino's satellite villages have a village nickname; *gatti* (cats) from Sant'Angelo in Colle, *orsi* (bears) from Castelnuovo dell'Abate, *streghe* (witches) from **Camigliano** and *granocchiai* (frogs) from Buonconvento. Tuscans are known in the rest of Italy as *mangiafagioli*, bean-eaters, because so much of the rustic cuisine (*cucina povera*) relies on legumes.

A prominent physical feature, a unique behavioral trait, an interesting habit or occupation, a place of origin, a favorite food, or a memorable incident can all result in nicknames. They are often highly ironic; local historian Bruno Bonucci tells of a 1617 reference to a Montalcino blacksmith known as *Farfallino* or little butterfly. Another historian, Stefano Cinelli Colombini, told me that there are records of the Caporali family of bakers nicknamed *Chiodo* (nail) in the 1600s. Their heir still goes by that name. Diminutive and augmentative **suffixes** are part and parcel of the tradition, and often a chance remark can engender a nickname for life. One example: in Sant'Angelo in Colle, a six-year-old child was walking on a wall and was admonished by his mother. He replied, 'I'm not *tonto*' – and then fell off. This man was known as *Tonto*, an informal word for dumb, for the following seventy years. Nicknames, once bestowed, are hard to escape.

In 2019, a slim volume was published in Montalcino as a fund-raising initiative for the new **school**. It was called *Dimmi il soprannome... ti dirò che è* (*Tell me the nickname and I'll tell you who it is*) and had five hundred Montalcino nicknames, with a space adjacent so that people could fill in the real identity of the people in question. At the time, Montalcino town can't have had more than three thousand inhabitants within the town walls, which means that at least one in every six people was well-known by their *soprannome*. To be fair, there are several people whom I know only by their nickname. Remember, here even statues get teased, and a beloved and respected mayor was hailed by his nickname on the street.

Not Invented by Leonardo da Vinci

Although his contribution to science, mathematics, engineering, anatomy, art and architecture is incontestable, da Vinci has been erroneously credited with a number of impressive inventions, including parachutes, scissors and the prototype of the **bicycle**. Importantly for the wine industry, he also did not invent **airlocks**.

Oidio
Powdery mildew can attack any green parts of the vine and particularly enjoys the combination of hot days and cool nights. It prefers dry conditions and is therefore known as 'the fair-weather fungus'.

A dense canopy is its ideal environment. A transparent web covers the young shoots, surfaces of the leaves and even the green, unripe berries themselves. After about two weeks, grey-white, flour-like spores appear. The leaves look like they have been dusted with flour or ash. The disease is also known as *calcare* or limescale.

Oinochoe
This Greek-inspired wine-pouring jug, with its narrow neck and thumb rest, was also favored by the Etruscans. It boasts a near-full set of vowels. There would be no need for such a jug if you were being abstemious, which does have all the vowels, in the correct order. Apologies for being facetious (once more, there they are, all lined up).

O'Keefe, Kerin
Kerin is a Massachusetts-born wine critic of Italian wines, who has written two brilliant books on Montalcino. She is also an expert on the wines of Piedmont.

Ombrone and Orcia
These rivers border the square-shaped base (each side approx. 15 km) of the Montalcino slopes, along with the Asso. For some growers, the proximity to the river mitigates the heat of summer nights.

Organoleptic
This refers to what can be understood via the senses, ie what you can see, smell and taste of a wine as opposed to the dry data from a laboratory analysis. It is a more mystical quality that defies precise definition since the organoleptic properties of a wine also stretch to the reactions and emotions that the wine arouses on drinking. **Brunello** meets with **DOCG** approval only if it meets both the parameters of a lab test and receives the unanimous approval of a human tasting panel that judges the wine on its taste, appearance and aroma.

Oro di Montalcino
Don't miss this wonderful museum complex. An entry ticket gives you access to the recently restored Sant'Agostino church and its fourteenth-century *Bartolo di Fredi* frescoes, and also affords entrance to the *Museo Archaeologico*, *Museo di Arte Sacra* and the ***Tempio del Brunello***. There are two beautiful cloisters and a well-stocked bookshop, including the ubiquitous ***Eroica*** garb.

Feeling peckish? We leave museum cafés and their carrot cakes to other cultures; here it's a massive chandelier, cold cuts and a glass of **red** in the *Enoteca Bistrot*. If you ask to use the bathroom, someone will tell you to nip behind the Madonna, a beautiful painting conveniently placed to hide the entry to the ground floor loos. Look out for the **Etruscan** artifacts from ***Poggio Civitella***, the original **Wolf Statue** and the Saint Sebastian by Andrea delle Robbia. Consider downloading the app and make sure to enjoy the virtual reality visors for a swooping sense of the area.

Ossi di Morto
Literally translated as Dead Man's Bones, these are, in fact, delicious biscuits that are made with egg whites and almonds. They are crunchy and light and are considered a specifically Montalcino delicacy, though they do actually exist in various forms throughout Italy.

OTBN

Open That Bottle Night was created by Dorothy J. Gaiter and John Brecher, a husband-and-wife team who wrote about wine in *The Wall Street Journal* from 1998 to 2009. They invented OTBN in 1999 and it has been celebrated all over the world every year since then. It is always the last Saturday in February and anyone who loves **Brunello** needs to have this date in their diary. We all have That Bottle; always too good for the company or for the food that is being served, bought somewhere special, lugged from home to home and needing a reason to be opened. OTBN enables us all.

P

Palazzo dei Priori

What trip to Montalcino is complete without a photo of the clock tower and of the view of the **Val d'Orcia** seen framed by the steep alleyway past the *Grappolo Blu* restaurant on the *Scale di Via Moglio*? Any visitor should have at least one of these images on their camera-rolls; I have several of mobs of tourists immortalizing precisely those two shots.

The bell – *Il Campanone* – dates from 1262 and still marks time for the **city**. The main square, **Piazza del Popolo**, is dominated by the tower, a piece of architecture more Florentine than Sienese and relatively unique in this respect when it comes to medieval municipal buildings. Elegant and imposing, the ancient town hall was a polyfunctional space: the *Palazzo del Podestà*, the tribunal, the prison and the town council. The severe stone construction was built in the thirteenth and fourteenth centuries, and at one point was where the exiled Republic of Siena minted their coins.

The loggia that houses **Gian Gastone** was added in the sixteenth century. The walls bear insignia and coats of arms, along with the votes cast for and against the Tuscan territories becoming part of the Kingdom of Italy in the 1860 referendum. Once the seat of the **Consorzio del Brunello**, it houses part of the tourist office, is where civil marriages take place, and where the vintage tiles or ***mattonelle*** are embedded. We call it the *Comune Vecchio*.

Panfilo dell'Oca

This **Borghetto** street is named for the chap who handed over the keys to Montalcino to the Florentines and Spaniards in 1559, or – depending on how you like your history – the brave warrior who in 1553 led the ***ilcinesi*** in their defense of liberty against the Spaniards. There is a monument to him at the water fountain, created by Sienese artist, Massimo Lippi.

Passo del Lume Spento

The evocatively named *Passo del Lume Spento*, which I like to translate as 'The Guttered-Candle Pass', is so called because of the **wind** that used to extinguish all the lamps on horse-drawn carriages. This is the highest stretch of Montalcino at 621 meters above sea level. The *rettilineo* (finally a straight bit of road… if you've driven it, you may remember this with relief) is a good vantage point for seeing a glimmer of sea in the far distance at sunset, and Corsica and Elba may be sighted on occasion.

Peronospera

Downy mildew is a fungal disease that can cause severe damage to the vine. The mildew penetrates multiple cells and steals nourishment at the expense of both flowers and fruit. Yellow, raised spots appear on the leaves and the underside is coated with white mold. The vine, deprived of nutrients, cannot mature the grapes and the berries dry out and become necrotic. The same thing can happen to the flowers in very wet springs. Growers control mildew by applying copper sulphate sprays or various forms of sulphur. In 2023, fifty cases were reported in Montalcino and damages were estimated at 2.5 million Euro.

pH

Like the Richter scale for earthquakes, pH is a logarithmic scale. A wine with a pH of 3 is ten times more acidic than a wine with a pH of 4. The higher the pH, the lower the **acidity** and vice versa. Counterintuitive, I know, but a wine with a pH of 3.8 is less acidic than one with a pH of 3.4. Although total acidity and pH are related, they represent different ways of measuring acidity of wine. pH is a measurement of the strength of acid, while total acidity is a measurement of the percent by weight of an acid. Total acidity affects the taste of the wine, whereas pH affects the stability. The ideal pH for **Brunello** is between 3.38 and 3.45. In challenging vintages like 2023, it might hit 3.60.

Phenological Stages

Phenology, spelled with an f in Italian and much beloved by Italian wine people, refers to the study and change of plant organisms during the growing year, in relation to climatic and environmental conditions. Phenological stages are the changes in vine development as they respond to external stimuli such as temperature, light and humidity. Think of it as the milestones associated with child development. There are phases in the life cycle of a vine: *il pianto*, when the vines begin to weep after pruning, **bud break**, flowering, fruit set (*allegagione*), *invaiatura,* ripening (*maturazione*) and, finally, *riposo invernale* (winter dormancy). A deep knowledge and a series of expectations regarding the phenological phases are what allows the *agronomo* to decide on the optimal timing of vineyard operations such as *potatura* (pruning), *cimatura* (topping) and of course, **harvest**. Phenology is a powerful tool for understanding the implications of the climate crisis on viticulture.

Pianello

Identified by blue and white flags, *Pianello* is one of the Montalcino *quartieri*. Its patron saint, San Pietro, is celebrated towards the end of May.

Piazza del Popolo

This is the main square in Montalcino, dominated by the clocktower and animated by the ***Fiaschetteria Italiana*** and the beautifully ornate *Farmacia Salvioni*. It has had many names: once the *Piazza del Mercato*, after the **Unification of Italy** it became *Piazza Principessa Margherita*. In 1878, when the Princess became Queen, the square was upgraded to *Piazza Regina Margherita*. In 1945, the government issued an edict that all street names that referred to the disgraced royal family should be cancelled and so it was re-christened *Piazza del Popolo*, the People's Square. In summer, sit on the marble steps of the **Cappellone** and you will see that it really is the people's square. If you stay long enough, you will see the whole town walk by. In Italy, the habit of walking up and down the main street, exchanging pleasantries, gazing and being gazed upon, is described as doing a *vasca*. The same phrase is used for completing a length of a swimming pool.

Piazza Padella
I had to research the real name of this square which turns out to be called *Piazza Garibaldi*. I'm not ashamed. *Padella* means saucepan, and is the shape of the square if you consider Via **Panfilo dell'Oca** to be the handle.

Pied de Cuve
Fermentation with wild **yeasts** can be bump-started with a *pied de cuve*, and those producers that choose to forgo this shortcut are to be admired for their courage and confidence in their vineyard practices. Essentially, a small batch of wine is prepared to cultivate a population of viable wild yeasts. This culture is then added to the rest of the grapes when they are picked, rather like a sourdough starter.

Pinci
Vowel changing and letter adding in words is terribly common here, in order to differentiate pronunciations even between small villages. No one who goes to Montalcino can avoid eating the ubiquitous *pinci* – egg-free, hand-rolled, irregular, fat spaghetti. This same pasta (which can only be found in Siena and Grossetto provinces), outside the walls of Montalcino are called *pici*.

Plums
In April, be sure to drive down to Sant'Angelo Scalo to admire the plum orchards in blossom. Down by the river, Banfi has one hundred **hectares** of Agen plums, a variety particularly suited for drying, and is the largest Italian producer of this variety of plum. In Italy, there is a clear distinction between *susine* and *prugne* which are actually different species, but in English we have to make do with the one word. This is another link with **Napa,** a city with which Montalcino was officially twinned in 2021.

Poggio
Pronounced pod-jo (just two syllables, not three) this word means gentle hill and is a synonym of *colle*. *Poggio* is a toponym in many regions but these days is only used in common speech in Tuscany. In other parts of Italy, *poggio* is relegated to poetical and literary usage. Some of the English translations – hillock, knoll – are in a similar category. In Montalcino, there are many *Poggios*, both on wine labels and landmarks, and quite a few *Colles* too.

Poggio Civitella
This is the only completely excavated and restored Etruscan fortress in the world, just three kilometers from Montalcino. It is at 660 meters above sea level, high above the **Passo del Lume Spento**, offering a strategic viewpoint both towards the Apennines and the Tyrrhenian Sea. A series of excavation campaigns, conducted between 1993 and 2005 by the University of Florence, exhumed the fortress that was built towards the end of the fourth century BCE on the ruins of a previous village. Many of the found items are on display at the Montalcino Museum. Unfortunately, the site is not currently open to the public, except on rare occasions

Poirot, Hercule
Multiple websites cite an unnamed Agatha Christie novel in which Poirot disproves an alibi by knowing that the culprit could not have tasted a 1900 **Brunello** since this vintage was not produced. The real mystery is if this was just invented by a clever marketer to snowball its way onto brochures and websites, or if the novel actually exists. The Christie Archive has settled this puzzling matter once and for all: sadly, there is no such plot twist in any of the novels.

Ponti
Ponti (bridges) are an integral part of Italian life and one of the first things that everyone checks when they get their free calendar from the butcher

or the banker at the start of a New Year. A *ponte* occurs when a national holiday falls on a weekday (ideally close to the weekend) and the whole country will hope for a day or two off work to get a long weekend. **Schools** are often closed on the intermediate days when Tuesday and Thursdays are involved. Since national holidays are on fixed dates there can be awful years when nearly all the red-circled days fall on the weekends (so no *ponti*, but extra pay if you are a salaried employee) and there can be *ponti*-rich years, with whole months (April and May particularly) riddled with half-weeks.

A good run of *ponti* can completely change the turnover of the hotel and restaurant trades, not to mention bump up the number of caravans and campers on the roads. Panicked news broadcasters talk about *esodi* and *contro-esodi* to describe the vast numbers of people going from A to B and back again. The critical dates are 25 April, 1 May, 8 May (just Montalcino, for the **Santo Patrono**), 2 June, 15 August, 1 November and 8 December. The full list includes 1 and 6 January, Easter Monday and 25 and 26 December. Add these to the minimum twenty days statutory holiday for workers and there is a total of around thirty-one days off per year. My favorites are the *festività soppresse,* national holidays that no longer identify as holiday but are still paid extra or can be switched with time off as a gesture to Italian history. There are four of them: *viva l'Italia!*

Porte

Montalcino has six different *porte,* gates or entry points to the town, though you may not be aware of passing through them; *Porta Castellana, Porta Burelli, Porta al Cassero, Porta Gattoli, Porta del Cornio* and *Porta Cerbaia.* According to the sixteenth-century historian, Domenico Cerratti, before the battle of Montaperti, six powerful Tuscan families committed to protecting Guelph Montalcino from Ghibelline Siena. This is the reason for the six hummocks at the base of the Montalcino insignia, blood red to represent the loss of life. This could easily be a legend, yet it would explain the names of the **city** gates.

Post Office

If you absolutely have to go in here, take a deep breath and abandon yourself to forces beyond your control, especially if it is pension day. In summer you could fry an egg on the cash machine, so remember to bring a parasol and gloves. And if it's just stamps you are after, go to a *Tabacchi*.

Potatura

Pruning, or *potatura*, is the moment in which the positioning, number and weight of future bunches of grapes is decided, along with the canopy vigor and leaf distribution of the vine. These two activities are regulated by the number of *gemme* and their position. Surgeon-like skill can be required, along with the ability to triage quickly. A skilled *potino* will be simultaneously evaluating the age and vigor of each individual vine, weighing bud viability, the length of the shoots, climate, **terroir**, rootstock and predicted weather conditions as they wield their scissors.

Pruning in Montalcino is performed when the vines are dormant, though cold snaps in late spring in recent vintages have pushed many wineries into pruning later than ever before. After *potatura* – the cutting – ***stralciatura*** has to happen. This is the act of dragging the woody shoots or canes (*tralci*) through the canopy wires and piling them up at the foot of the vine-rows to be burned or buried. A green harvest, *potatura verde* is *de rigueur* in the late spring to contain and manage the canopy and/or to improve microclimatic conditions around the bunches and to improve the penetration of any airborne 'treatments'.

Presepi

Presepi, or nativity scenes, are common in most Italian homes. They are set up on 8 December and dismantled on 6 January. When we lived in Montalcino, Marco and the children set ours up in the pizza oven, the potential blasphemy seeming to escape nearly everyone. Now, in our house in Siena, we have it in a bookcase: eminently more conventional.

In Montalcino, an extraordinary display is created by **La Titina** in the gardens of **Il Prato**. As is common in Italy, the scene is not restricted to the manger, the main players and the three kings. There are water carriers and fishermen, bridges and wells, and a desert scene with camels, along with a whole village going about their tasks. Each year, there are additions and changes: an ambitious palm-tree oasis or home-made ladders for all the miniature olive trees.

Pump-over
Pump-over (the re-circulation of wine from the bottom of the fermentation vat to the top to soak the grape solids) is sometimes used to aerate the wine but does not provide the same effects as ***délestage***, because the wine is never separated entirely from the grape solids.

Pungitopo
This plant (*Ruscus aculeatus*) with very spiky leaves and bright red berries, can be seen in the Montalcino ***bosco*** in the winter months. It is known as *pungitopo* – or mouse-pricker – since it used to be hung from the rafters to protect *prosciutti* from mice. No Christmas would be complete without an excursion to pick some in the woods to decorate the home. Holly (known as *agrifoglio*) exists too but *pungitopo* is much easier to find, though devilishly difficult to transport and arrange.

Punt
Nothing to do with betting, kicking or messing about in boats. The punt is the name for the large indent at the base of wine **bottles**, the ideal place for a thumb when pouring one-handed, your other arm bent, wrist resting in the small of your back. This allows for service at arm's length, always nicer for guests, as well as avoiding covering the label, gripping the neck or risking heating up the wine with overly warm hands.

Of course, all wine bottles were once hand-blown. The base of any hand-blown vessel starts as the end of a glass bubble. The blown bubble has to be taken off the blow pipe and put on a long iron rod which is attached to the base of the hot bottle while finishing in order to create straighter sides. This rod is called a pontil rod, and is also known as a ponty, punte or punty, hence the name punt. Strangely, in Italian this part of the bottle is just called the *fondo* or bottom of the bottle. Instead of a rounded or pointed bottom, glass-blowers pushed the seam inwards, creating the concave dimple that those of us with strong forearms know and love. The presence of the punt originally meant that bottles had a chance of standing upright and provided structural integrity, particularly for wines with bubbles or secondary fermentation in the bottle. There is some disagreement as to whether or not the punts assist in collecting sediment, particularly given modern winemaking methods, or whether they are a clever ploy to deceive consumers regarding bottle volume. Certainly, for many years, punt depth and bottle weight were both considered indicators of quality and prestige.

Quartieri

Montalcino is divided into... you guessed it, four quarters: **Borghetto, Pianello, Ruga** and **Travaglio**. Each has its own church, headquarters, emblem, colors, songs, town notice board for birth announcements and, most importantly, **archery** team and practice area. Children are baptised into their quarter of birth and the headquarters are often the location for birthday parties.

Like the *contrade* in Siena, the *quartieri* provide an important social structure, whether it be dispensing **school** bursaries, organizing traditional holiday activities or increasingly elaborate celebrations for **Sagra del Tordo** victories which might involve turning a steep urban descent into a ski-slope for a day or creating a seaside resort in the middle of town. These things have to be seen to be believed. It is not uncommon to see *quartiere* tattoos and signet rings, and no Montalcino child goes without the opportunity to be 'dressed' for the pageants during the *Sagra del Tordo* or the **Apertura delle Cacce**. A Montalcino couple's goal is to be the Lord and Lady of one's own quarter. An indication of the importance of the *Sagra* could be the fact that the week before both festivals is known in Montalcino as the '*settimana santa*' – which most Italians would associate with the week before Easter, possibly the most sacred moment in the Catholic calendar.

Quercetina

The world expert on *quercetina* in Montalcino is Italy's first Master of Wine, Montalcino's own **Gabriele Gorelli**. This was the subject of his thesis, which can be downloaded from the MW site for deeper reading. Bottom line: *quercetina* is an increasing issue in Montalcino due to the ever-hotter summers and was first recorded as an issue here in 2002. It is a polyphenol that the vine produces in response to light exposure for protection. Although it has multiple health benefits, in wine it causes unsightly precipitation, which, although not affecting taste, is something consumers prefer not to see in the **bottle**, and in recent years it has been added to the list of things growers worry about prior to bottling. Recent research points to *quercetina* as being a possible cause of red-wine headaches.

Quintale

Un quintale – in English, a quintal – is equivalent to 100kg and is the unit used for grape **yields**. Although abandoned by the EEC in 1990, its use remains widespread in Italian and European agriculture. The word dates from the sixteenth century, from Spanish, via Arabic and Greek. Abbreviated to QL.

Q-words

There are only 485 words in Italian that start with the letter Q and now you know three of them.

Raffaelli, Ilio

Known to all by his **nickname**, *Bronzino*, Raffaelli (1926-2023) was a revolutionary mayor of Montalcino from 1960 to 1980, crucial years for the development of the town. He had an overriding vision of Montalcino as a *città di primati*, a **city** of world firsts, and during his tenure, Montalcino's turnover increased by 50%. He is considered by many to be responsible for ***Brunello's*** fame worldwide, due to his great and bold decisions to elevate *Brunello* and promote existing local resources rather than authorizing the construction of factories.

Son and grandson of woodsmen, armed only with a primary school diploma, he was a pragmatic and free-thinking man who published many volumes about Montalcino. He was a vocal supporter of sustainable agriculture, a differentiated economy for Montalcino and the return of a functioning **hospita**l.

He died in 2023, just short of his ninety-seventh birthday, reading three newspapers a day until the end. He worked as a *carbonaio* until he became a member of the *Partito Comunista* in 1945, and was both a soldier and a *partigiano* in the Second World War.

Rainfall

These days, it's hard to identify average rainfall – just think of the **drought** of 2017 or the legendary rainy vintages of 2002 and 2014. More recently, in 2023 a wet May was a real challenge for growers. Tuscany was spared the terrible flooding that devastated Emilia Romagna, but received over 220mm of rainfall, nearly five times more than usual in that period. ***Val d'Orcia*** was amazingly lush and verdant. Everything, vines included, grew at a rate of knots. The almost daily afternoon rains meant mud and restricted tractor access and an increased risk of mold and mildew. In summer, we can easily have two consecutive months with negligible rain and temperatures between 33°C and 40°C during the day.

Recinzioni

Recinzioni are entrenched fences or electrified enclosures around the vineyards. They have become increasingly necessary due to the relatively unregulated deer and boar populations in Tuscany. Controlling perimeters is an important part of vineyard management, particularly in August, since a fenced-in boar family is a dreadful scenario for a **winemaker**.

I was shocked to learn that fencing off vineyards is not permitted in San Gimignano, or at least not within certain proximity to the town walls. The landscape in Montalcino has been uglified in recent years by the inevitable introduction of vineyard fencing but it is the only way to protect precious fruit from the dreaded **ungulates**.

Red

Tilt the glass gently to the side to form a perfect oval. ***Sangiovese*** has a wonderful light-catching quality to it which in Italian would be described with the words *limpido* and *trasparente*. Obviously, it is far from colorless, but compared to some inkier grapes, it has a certain brilliance. You should be able to read text through a glass of **Brunello**. Young *Brunello* might start out a brilliant ruby with purplish hues to become a bricky orange with age. Be familiar with the whole panoply of reds; tonalities and reflections can range from almandine, garnet, cinnabar and vermilion to the more abstruse jarosite and hematite. Some consider it a success if you need to google their tasting notes, so do your homework.

Red Montalcino

A promotional event for ***Rosso di Montalcino*** held in June or July, which took place for the first time in 2022. There is food and music, and the aim is 'to promote the contemporary soul of an amazing territory'.

Relationship with Siena

Historically, Montalcino has had a fraught relationship with its close neighbor, Siena, just 41km away. Although much of the dominion dates back five hundred years or so, even today, no love is lost between the two. The story of the **Wolf Statue** is very telling.

Resa

According to **DOCG** regulations, the average **yield** per vine cannot exceed 2.7kg. Depending on the vine **density**, yields of grapes per **hectare** cannot be more than eight thousand kilograms (although this is often abandoned in challenging vintages such as the very hot vintages of 2017 or 2021). The yield in wine cannot be more than 68%.

In France, it is common to refer to the yield in wine per hectare but in Montalcino we refer to the yield in grapes. Due to *Sangiovese's* great variation in grape size, skin to flesh ratio and so on, the yield in wine can vary wildly from vintage to vintage. Low yields are synonymous with high quality in fine wines, so many wineries prune to obtain far less than the eight thousand kg allowed. That said, the rise of *Guyot* as a **growth system** is part of a move towards higher yields per hectare.

Riserva

What makes a *Brunello* a *Riserva*? The **DOCG** regulation requirements are minimal; the wine need only be released a year later, with an additional two months of **bottle**-aging (six months versus four months). What is to stop producers just re-marketing unsold vintage *Brunello* a year later as *Riserva*? To prevent precisely this, a producer's intentions must be formalized prior to the release of the vintage.

In reality, *Riserva* may not be in constant production. Some wineries only commit in years of exceptional quality. Others make it from a particular vineyard, or a particular selection of grapes that has a different wood path. One producer, Le Chiuse, keeps their *Riserva* for ten years before release. The first *Brunello* ever made was the *Biondi Santi* 1888 Brunello Riserva.

Rope-making

Until the end of the Second World War, most farming families in the *Val d'Orcia* were cultivating hemp for personal use to make clothes, ropes, sacks and so on. In Montalcino, up until the mid 1950s, below the walls of *Viale Strozzi (Viale della Madonna)*, a rope-maker had an outdoor workshop where ropes and sails were produced. If you look carefully, you can see the grooves worn in the **city** walls. Here the ropes used in the Palio in Siena were made, twelve and fourteen meters long, each weighing around fifteen kg.

Roses

If you are driving around Montalcino or visiting wineries, it is impossible not to note the roses that are planted at the end of the vine-rows: bright nail-polish pink, hazy yellows and ruby reds. Historically, farmers relied on roses as a sentinel, rather like a canary in a coal mine. Susceptible to many of the same diseases as the vineyard, roses provided advance warning for necessary intervention, for example, spraying to protect against *oidio*. It is true that their function is primarily aesthetic these days, and it is rare to find them anywhere other than at the traveled parts of a vineyard, but they are still very beautiful.

Rosso di Montalcino DOC

Lovers of Italian wine will be familiar with the idea of *Rosso di Something* wines that are either connected to a place or to a superior appellation. In Italian, these wines are described as *DOC di ricaduta* or drop-down **denominations** from the 'parent' **DOCG**. *Rosso di Montalcino* was the first **DOC** of this kind and was codified in 1983, just three short years after **Brunello** was elevated to DOCG status. Actually, its pedigree goes back much further. In the late 1800s, a wine called '*Rosso*' was produced in Montalcino, and even won a bronze medal in 1869. In the 1960s, *Rosso* was a DO wine, and wines from this time were described as '*Rosso dai Vigneti di Brunello*' which then became *Rosso di Montalcino*. To begin with, there was a separate *Consorzio* for *Rosso*. As far as the disciplinary is concerned, grapes must be sourced in Montalcino and must, of course, be 100% **Sangiovese**. The maximum **yields** are higher than permitted for **Brunello** (nine thousand kilograms per **hectare**, and 70% in juice), the **ABV** is lower (12%) as is the **acidity** (minimum 4.5g per liter versus 5 for *Brunello*). While it must be bottled on Montalcino territory, screwcaps are permitted. I almost fainted just writing that and have never seen one in use.

Rosso di Montalcino can be released from 1 September following the harvest year and has no obligatory time in wood. Anyone with an over-performing elder sibling will already sympathize with *Rosso's* plight, often condemned with phrases such as *Baby Brunello* or junior partner.

In 2022, there was a big institutional push to give *Rosso* its own share in the spotlight and its own party, rather like **Benvenuto Brunello**, called **Red Montalcino**. The buzzwords for this new vision for Brunello include joy, versatility and contemporary, and the hope is that this younger wine will be attractive to younger generations of drinkers.

It used to be that there were vineyards authorized for *Rosso* or *Brunello*, but now a quota system is used. There is a bewildering variety in the *Rosso* 'offer' and it can be true that sometimes *Rosso* are made for reasons dictated by cash flow and/or cellar space or by *Brunello*-blending decisions. It is conceived to be lighter than *Brunello*, both on the pocket and on the palate. A good *Rosso* is a thing of beauty and can give incredible insight into the vintage. On occasion, a great *Rosso* can even eclipse a middling *Brunello*.

No need for a crystal ball to predict that we are going to be seeing a lot more *Rosso di Montalcino* in the future. In December 2023, production was given an enormous boost by the producers' vote to almost double the vineyards authorized for *Rosso*. The land for *Rosso* was increased to 883.7 hectares by adding 364 hectares to an existing 519.7 hectares. This meant that average bottle production was increased from 3.6 million to over 6 million. The decision, ratified by the *Consorzio*, allows smaller wineries to increase their lands by 15%, whereas larger wineries have a sliding scale of lesser percentages based on dimension. It should be clarified that the 'extra' vineyards already existed but previously belonged to other **denominations**. This means that there will be no change to the landscape due to large-scale planting, nor waiting for new vineyards to become productive. The increase in production should kick in from harvest 2024.

Ruga

One of the Montalcino *quartieri*, identified by the colors blue and yellow. On the last Sunday of July, *Ruga* celebrates its patron Saint, San Salvatore, a Franciscan martyr who was born in Montalcino.

Sagra del Tordo

The *Sagra del Tordo*, revived and formalized in 1957, is separated into two major events. The *Sagra del Tordo*, the Festival of the Thrush, takes place on the last Sunday of October; the **Apertura delle Cacce**, the Opening of the Hunting Season, is always the second Sunday of August. Not just an excuse to overuse #menintights, over a hundred Montalcinesi dress up in beautiful medieval costumes and participate in the *corteo storico*, or pageant. It is very odd to see familiar faces so transformed: is that the butcher in those stripy tights? Could that be the post-lady with a twisty globe on her head? There is chanting, drumming and much quaffing of local wines. Visitors can eat food prepared by each **quartiere,** served by the youngsters.

Heartfelt as it is, the *Sagra* is also a source of revenue for the whole town, the last 'good' weekend of the season for businesses and for funding the *quartiere*'s renovations, study grants and annual activities. Winning the *Freccia d'Argento* is very important; tears are shed, teenagers shriek and everyone gets very tense. For weeks afterwards there are in-jokes and teasing, with strange items appearing overnight under the **Cappellone,** the vaulted space on the main street. Victories and losses are never forgotten since each *quartiere* keeps an ongoing tally.

San Giovanni d'Asso

You might be forgiven for thinking that 2016 was the year of the Brexit referendum but in Montalcino an equally important subject was being put to the public vote. The residents of San Giovanni d'Asso and Montalcino voted on whether to combine their two towns, joining forces to improve municipal services, reduce costs and increase visibility, although at the risk of subjugation for the smaller San Giovanni. This fusion was even the subject of a *New York Times* article, which described it as a 'municipal marriage of convenience'.

In 2014, the Italian government began to incentivize town mergers while penalizing towns with under five thousand inhabitants. As well as the perfect combination of truffles and **Brunello**, both small communities have had tangible gains from the merger. As the bigger partner, Montalcino insisted on keeping the name and the coat of arms. If you can, stop by in San Giovanni d'Asso at the wonderful contemporary sculpture park, **Il Bosco della Ragnaia**, created in 1996 by American artist Sheppard Craige, or the interactive **Tartufo** museum. Do not miss the exquisite *borgo* of Montisi.

Sangiovese

Some subjects cannot be reduced into bite-sized nuggets and the complex ampelographic history of *Sangiovese* is one of these subjects. I defer to experts Kerin **O'Keefe** and Ian D'Agata, particularly regarding the historical perpetuation and snafu of *Sangiovese Piccolo* and *Grosso*.

Sangiovese is a terrifically common Italian grape and is present in many **DOCGs** and **DOCs** as a mandatory component grape. It is, actually, the most cultivated grape in Italy, comprising more than ten per cent of Italy's vineyards. Argentina is second in *Sangiovese* production but grows less than half of the amount that Italy does. That said, the only wines in Italy that must be 100% *Sangiovese* – *sangiovese in purezza* – are **Brunello di Montalcino DOCG** and *Rosso di Montalcino* **DOC**

In addition to huge clonal variation, *Sangiovese* is a **terroir**-driven, genetically unstable, chameleon grape and consequently, depending on where it is grown, it presents differently. As a result, it now goes by many names: there are thirty known synonyms in Tuscany. Many of these grapes were historically considered separate varieties but have since proved to be all versions of *Sangiovese*. Here we call it **Brunello**.

Heads up: *Sangiovese* is ubiquitous, yet performs differently everywhere. Although it ripens late, it does not love the sun. It's tricky, temperamental and demanding – and in Montalcino it finds one of its best expressions.

Sant'Antimo
This gorgeous abbey is located near Castelnuovo dell'Abate. The first documentary evidence dates from 814 and there has never been any doubt about how to pronounce it: emphasis on the An if you please. Legend has it that the abbey was founded by Charlemagne (748-814 CE) to give thanks when his court and army survived the plague.

The church was rebuilt in the eleventh century and is one of relatively few Romanesque churches in Tuscany. It is made from beautiful stone, quarried locally. The simplicity may be a relief if you have had your fill of stripy, wedding-cake churches. Enjoy the **cypress**-lined approach, the lone cypress as tall as the belltower, architectural details, pleasing symmetries and spiritual ambience. Look out for the Roman cornucopia in the side wall, allegedly pillaged from a Roman villa. Inside, locate the capital depicting Daniel in the lion's den by twelfth-century French sculptor, The Master of Cabestany. I am always amused by the small informational sign forbidding marriages via a wordless image. Only residents of Castelnuovo dell'Abate can be wed in this sacred space.

If you have a weakness for tinctures, honey and soaps prepared by members of religious orders (Trappist chocolate, anyone?), then you will enjoy the shop. Be prepared to be fined if you do not pay for a parking ticket; not very Christian, I know, but it happens rather often.

Sant'Antimo DOC
This baggy and catholic **denomination** joined the ranks of the Italian **DOC**s in 1996 and is pretty much based on the concept of 'anything goes'. The only real rules are that grapes come from the designated Montalcino territory, the wines must be bottled in the province of Siena and that **yields** in grapes do not exceed 9,000 kg per hectare.

Sant'Antimo can be white, **red**, *Vinsanto, Vinsanto Occhio di Pernice* (which includes a percentage of red grapes and is named for its tawny color, just like that of a partridge iris), *Novello* and also as a varietal wine, ie composed primarily of the named variety (in this case minimum 85%):

Chardonnay, Sauvignon, Pinot Grigio, Pinot Nero, Cabernet Sauvignon and Merlot. *Sant'Antimo DOC* was created in response to the limitations on land authorized for the production of **Brunello** and to the demand and interest in international varieties, fueled by the rise of the so-called **Super Tuscans**. The 2023 increase in *Rosso di Montalcino* vineyards will undoubtedly lead to a decrease in production of *Sant'Antimo*.

Santo Patrono

All Italian towns have a *Santo Patrono*, a saint's day and it is a paid holiday for all workers, as per legislation passed in 1949.

In 1913, Montalcino officially claimed *Maria Santissima del Soccorso*, Our Lady of Succor, as the town's patron saint with celebrations every year on 8 May. In 1553 during the eighty-one-day siege of Montalcino, the Virgin appeared over the **Fortezza** and the siege (*assedio*) was interrupted.

The church dedicated to *S. Soccorso* stands alone, looking out over the **city** walls towards Siena. It was erected across the centuries and combines all sorts of styles: a renaissance layout, baroque interiors, and a neoclassical façade. In Montalcino parlance, **La Madonna** refers to that whole stretch of the city walls and is a common walk.

The 8 May celebrations are, of course, largely religious, and include the blessing of ambulances in the main square, a town-wide bingo session and a firework display.

Scheletro

If you are poring over technical sheets for **Brunello** you are bound to come across this word in conjunction with soil. *Scheletro* translates in English to skeleton, which can be confusing. Its presence means that the soil texture contains fragments of material that are greater than 2mm in diameter. The word comes via the Greek, meaning arid. This sort of fine gravel is ideal for making great wines because it does not retain water and is very permeable to roots.

School

Montalcino had the first full-time (ie afternoons too) primary school in Italy. It opened in 1970. This national innovation was the subject of a TV documentary.

Three things to know about secondary education in Italy:

The first is that children need to identify a direction at the age of fourteen, to choose what kind of secondary school they attend for the next five years. This decision is based on aspirations, academic ability and, often, what schools are within an acceptable distance.

The second is that schools are divided into three main types. One, a *liceo* where Greek and Latin may be taught; studies, even in a scientific *liceo*, are largely theoretical and lead inevitably to a university degree. Two, the *Istituto Tecnico Professionale* – a school with a specialty, eg tourism or economics. These institutions tend to be more practical and the nineteen-year-old who exits them is already trained for a profession and the world of work. Finally, there are the ITS, the *Istituti Tecnici Superiori*, which train for specific jobs such as plumbers, electricians or beauticians.

The last thing to know is that it is culturally normal for kids from rural communities to spend as much as two hours each way commuting to the school of their choice and for bleary-eyed parents to be taking them to the bus-stop for 6am departures. The reason it takes so long is that school buses take a roundabout route that stops at all the villages along the way.

Montalcino has two secondary schools, a *Liceo Linguistico* which has been around since 1931, specializing in languages and tourism and an *Istituto Professionale Agrario* that welcomed its first students in 2017. This gave a generation of potential agronomers and their parents an extra hour of sleep and was a brilliant educational option for this town. The Montalcino school is a branch of an established and esteemed Sienese institution named after the second Count of Brolio, Bettino Ricasoli. The natural path to study winemaking or vineyard husbandry is to attend a school of this kind before further university education.

Scirocco

Scirocco – remember *sci* is pronounced *she* in Italian – is a warm **wind** that comes from the Sahara, picking up moisture from the sea *en route*. It intensifies the heat of the summer and puts everyone in a terrible mood. When it occurs for an extended time, it is feared by growers because it increases air humidity and the risk of **oidio.** If someone calls you a *sciroccato*, it's not exactly a compliment – it means you are bizarre and unpredictable.

Settimana del Miele

This honey festival was inaugurated in 1976 and, although described as a week, has shrunk to a three-day affair in recent years. It is what is known as a *mostra mercato* and you can purchase honey and related products: candles and cough sweets, hives and bee-keeping outfits. It takes place in early September.

Sexual Confusion

This is one way to flummox insects that affect the vineyards, specifically the European grapevine moth *(lobesia botrana*, known as *tignoletta dell'uva)*. Strange, brown, plastic containers that look rather like chocolate bars are hung amongst the vines. They contain and emit synthetic pheromones identical to the natural pheromones released by the female grapevine moths. The males are befuddled and can no longer identify the presence of the female moths and consequently fertilization is prevented. The idea is that the moth populations will dwindle, generation after generation, season after season, because their natural mating patterns have been hindered.

Sfarfallamento

This beautiful verb describes the precise moment in which, after **bud break**, the vine leaves start to unfurl and the first leaves separate from the shoot tip. It comes from the botanical term for butterflies emerging from the cocoon, and it happens that the first leaves of **Sangiovese** do indeed also look like butterflies perching on the vine.

Sfemminellatura

Always hard not to be offended by the fact that the premature and useless suckers that appear in summer and syphon strength from the vine are known as *femminelle* (the girls). The operation of *sfemminellatura* is the removal of said shoots and is an obligatory manual operation.

Sfuso

This is loose wine, or rather wine in its unbottled, unfettered state. The Siena Chamber of Commerce issues weekly pricing as a guide to those who buy and sell **Brunello** before bottling. The closer to release, the higher the price. In 2019, 2014 *Brunello*, an indifferent vintage but one in short supply, peaked at €1,400 per hectoliter. Times have changed; in January 2024, 2019 *Brunello* is listed at €850-€1,000 per HL. The list is easily consulted: just use *camera di commercio listino Brunello* as your search term and then plough through the eight or so pages of priced items. The wines are usually on page four.

Slavonian Oak

Traditional barrels for **Brunello** are made from oak from Slavonia, a forest area of north-eastern Croatia – not to be confused with Slovenia, an entirely different country. Careful with those vowels now. The beautiful Slavonian forests are home to the largest and oldest oak tree in Europe dating from 1675. The climate is perfect and the **altitude** encourages optimal slow and even growth. The trees are grown closely together to create dense vegetation, promoting upward growth. The wood is prized because of the fine grain and scarcity of knots. The forests are managed sustainably, which is fortunate, given that, on average, two barrels are made from decades-old trees. Slavonian oak is suited for medium to big barrels (***botti***). It is gentle and consistent over time, giving long micro-oxygenation and extended cellarability, allowing the fruit to sing rather than endowing overly *boisèe* notes.

Snow

Yes, it does; most years, actually. Since the ploughs and trucks come from Siena, even a very small amount of snow can completely incapacitate Montalcino, given how steep many of its roads are. The first flakes are met with hurried evacuations and a mounting sense of panic. Shop grilles are pulled down, **schools** are closed and offices empty. It took me a while to get used to these extreme reactions, having been brought up in a part of the world where there are salt-bins on most streets and a few centimeters of snow pose no threat to the logistics of daily life. Here in Montalcino, the last heavy snowfalls were 27 February 2018, 12 February 2010 and 2 February 2012 (50 cm and temperatures at -10°C / 14°F).

In 2012, we were living at the winery and found ourselves trapped on the property, since our only entry was via dirt roads that had iced up beneath the snow. The council does not clean these secondary roads and our tractors could not climb them. The water pipes froze, we lost power and were melting snow for cooking: altogether too Laura Ingalls Wilder for my liking, particularly with three small children.

There is a proverb, *sotto la neve pane, sotto l'acqua fame*, which everyone likes to wheel out whenever it snows. Literally, it can be translated that there is bread under the snow, and hunger under the rain. Snow is good for grain crops, whereas rain can be damaging. The same is true for vineyards. Snow is an excellent way for the vineyards to absorb water with no top-soil or nutrient loss and this can be a boon if there is a dry summer. An added plus: sub-zero temperatures are an excellent natural insecticide, wonderful for both vines and olive trees. Sadly, the **cypress** tree does not enjoy snow; it makes the branches splay and ruins the sleek silhouette.

Solforosa

Sulphur dioxide is used as an antioxidant and an antiseptic. SO2 is judiciously used in the cellar, from fermentation to bottling, whenever a wine comes into contact with oxygen, eg **decanting**, filtering, bottling and so on. Part of the gas combines with some components of the wine while the remainder stays free, ie not combined, and able to do its job. The amount of free sulphur dioxide added to the combined amount determines the amount of total sulphur dioxide. As far as the European Union is concerned, the maximum allowed is limited to 160 mg/l for red wines and 210 mg/l for white and rosé wines. In the US the maximum legal limit is 350 mg/l. A good **Brunello** would usually have around 50 mg/l sulfites. From 2005 onwards, wines produced in the EEC have to indicate that a wine contains sulfites on the label but not how much.

Sorteggio

A very exciting moment, prior to the **Sagra del Tordo** and the **Apertura delle Cacce**, when the archers are selected for each *quartiere*. Three names of the most promising archers are submitted but only two are selected, plucked from a medieval bingo cage by a man in tights. It is my favorite moment of the two events and I can highly recommend it. Even if you are confused by the ceremony and can't remember the color schemes, the sight of each quarter surging into its designated area in **Piazza del Popolo**, the absolute silence that descends over the square, the palpable tension and the strong reactions to the selection need no translation. The competitive lusty singing that goes on has to be heard to be believed. Goose bumps!

Sovescio

Many wineries plant leguminous cover crops at different depths below the surface, depending on type of seeds and desired effect. Lupins and fava beans enrich the soil when ploughed and turned over in the following months. These plants have root nodules containing symbiotic bacteria able to fix nitrogen. Brassicas, like bright yellow mustard, act as a natural biofumigant, increase soil friability and suppress weeds.

The etymology of the term is uncertain but most philologists speculate that it is derived from the Latin *subversiare* meaning to turn under. As Italian wine expert Do Bianchi explores, on his eponymous blog, the practice of planting cover crops to replenish nitrogen levels in farmland soil dates back to Roman times. I reproduce his research here, with permission:

" 'First consideration belongs to the lupine [a legume]; wrote Columella in the first century C.E. (*De re rustica*), 'as it requires the least labour, costs least, and of all crops that are sown is most beneficial to the land. For it affords an excellent fertiliser for worn-out vineyards and ploughlands; it flourishes even in exhausted soil.' But the practice of planting cover crops as a means to improve soil 'health' became a hot button issue in early nineteenth-century Italy, when Piedmontese chemist Giovanni Antonio Giobert published his revolutionary research on *sovescio* and its farmland application. His experiments centered on the use of rye for green manuring (crop rotation was another focus). His work was met with unbridled disbelief in some quarters. After his greatest detractor, Count Carlo Verri, issued his response refuting Giobert's findings, one of their contemporaries noted that their dialectic represented the dawn of a 'new era' in Italian agriculture. Ultimately, Giobert's theories were embraced by hundreds of Italian farmers according to the anonymous author of an 1820 report on Verri's polemic."

Spampanamento

This is leaf-stripping in the days or weeks before harvest, in order to expose the maturing bunches to as much sun and ventilation as possible. If you crouch, you can see right through a vineyard. *Spampanamento* can be a high-risk practice insofar as the free-hanging grapes lose all protection from adverse weather. It can only be performed by hand. *Pampano* (the accent is on the first pa) is a fourteenth-century word from the Latin *pampinus*. In the rest of Italy, the word for vineleaf is *pampino*.

Naturally, for a wine-producing country, there are all sorts of proverbs that feature vine leaves: *'assai pampini e poca uva'* – all talk and no trousers – and *'dare pampini per uva*,' swindling someone by giving them leaves instead of grapes. But in Montalcino, just as we like an N in our **pinci**, they are *pampani*. In many Italian words, the S prefix flips a word into negative. We have *fiducia* (faith) and *sfiducia* (bad faith), *fortuna* (luck) and *sfortuna* (bad luck); here the S indicates removal, de-leafing in English, just as ***stralciatura*** means de-branching.

Spoofulated
Who can resist this delightful word? Spoofulated wines are considered over-extracted and over-manipulated, and rely on various (legal) winemaking additions or subtractions. A spoofy or spoofed wine has lost its sense of place, poor thing. There is no Italian translation.

Strada Bianca
Beware the white road, my friend, the wheels that skid, the clouds of dust! When a winery mentions in their Directions section that the last leg of the road to reach them is 'unpaved', *strada bianca* or *sterrato*, be prepared for the impression of driving along a desiccated river bed, and fervently hope that you will not encounter someone coming the other way at speed. Many of Montalcino's roads were only asphalted in the 1980s and the peripheral white roads that remain are often maintained by all the landowners along that route, which can be a cause of friction due to the complicated calculations of how many entry points and how much road leads to their property. If you have ever lived in a flat and have had to fight about who washes which stairs you will know what I am talking about. But, back to the white roads: just imagine the wear and tear for the ambulances and **school** buses. I remember being flabbergasted when I learned that Montalcino's school buses covered an astounding 600km a day (372 miles) when there used to be five routes in operation. This number has now dropped to three.

Strade Bianche
Strade Bianche is a road **bicycle** race that starts and finishes in Siena. First held in 2007, it is raced annually on the first or second Saturday of March. It is named for the white gravel roads in the *Crete Senesi*, which are a defining feature of the race. It often goes around Montalcino, meaning that you can be blocked for hours or find yourself an unwitting participant.

Stralciatura
This is the process by which the *tralci*, the previous year's woody shoots or canes, are removed after **potatura.** This is a punishing arm workout. I discovered, via a 1760 dictionary by Joseph Barretti, that the word *tralcio* also means umbilical cord (or navel string): a fact I will never need to know, but that will come to mind every time I think of debranching the vines. *Tralci* are heaped and often burned; the hills of Montalcino in January and February are dotted with long, straight plumes of white smoke.

Stress Idrico
This is what a parched vine experiences in the absence of rain.

Subzones
The politics and probability of Montalcino's subzones ever being mapped is a contentious issue, and very well-documented elsewhere. However, no resident of this area can deny the very real differences in soil, **exposition** and **altitude** of different parts of Montalcino and the consequent effect on **Sangiovese**, a grape renowned for being **terroir**-driven. There are warm pockets, where fruit trees flower first and the earliest wild asparagus can be foraged, and areas where vegetation tends to be developmentally delayed.

The strict legislation about land under vine has meant that many wineries own vineyards in different areas on the slopes of Montalcino, so most producers have firsthand experience of subzone variations and many make it part of their winemaking philosophy. It is a way of having one's eggs in multiple baskets should there be adverse weather conditions, and of staggering vineyard intervention and harvest.

Suffixes
In Italian, nouns can be transformed and supercharged by the addition of suffixes. A day, *un giorno*, can become a *giornataccia* in the time it takes to drink an espresso. The main suffixes that should be part of any visitor's artillery are: *ino* (small), *one* (large), *accio* (terrible, think of **vinaccia**), *etto*

(small and sweet). In Montalcino, an easy meeting point is the *Cappellone*, the large, covered, arched area in the center of town.

If you sit a moment to watch the world go by, you will no doubt hear the *Campanone*, the large bell on the tower of *Palazzo dei Priori*. Another example is the Tuscan dialect suffix *–aia* in the names of **Super Tuscans**: *Sassicaia, Ornellaia, Rondinaia* and *Solaia*.

Sughero

Sughero translates as cork. Natural closures are obligatory for ***Brunello di Montalcino*** and the use of agglomerated cork is forbidden. The rules are less stringent for ***Rosso di Montalcino.***

Superstition

I live with an Italian who considers any spontaneous well-wishing a form of *gufamento*. This word comes from the Italian for owl, the bird of bad omen or *malaugurio*. So now I check myself from inadvertently expressing good wishes for a football match or a day's fishing, since this is considered to cast irreversible bad luck on either experience and renders me directly responsible.

This sort of behavior is known as *scaramantico* in Italian. The word presumably derives from the Greek for chiromancy and is a form of superstition in which some phrases or behaviors attract or ward off good or bad luck: apotropaic in English.

The saying *Né di venere, né di marte non si sposa né si parte, né si dà principio all'arte*, which has its roots in pagan Rome, has governed some rather important life choices for our family, and we are not unusual in this. Translated, it advises the avoidance of doing important things on Tuesdays and Fridays: specifically journeys, marriages or the starting of artistic enterprises. Knowing this, it came as no surprise to me to hear of a Montalcino producer picking a symbolic crate of grapes on a Monday, so as to be able to schedule harvest proper for Tuesday.

Super Tuscan

This is how I describe my husband, but the term is more generally understood to refer to a style of wines that emerged on the Tuscan coast and in Chianti in the 1970s. *Sassicaia* vintage 1968 was released in 1971 by Tenuta San Guido in the village of Bolgheri, followed by *Tignanello* vintage 1971 from Antinori, commercially available in 1974. These iconic **reds** and their bedfellows (*Masseto, Ornellaia* and so on) command high prices all over the world. They were often originally labelled *Vino da Tavola*, the lowest of the low as far as **denominations** go, since they did not conform to any existing **DOC** regulations. French grape varieties were used; for example, Cabernet Sauvignon was blended with Merlot and **Sangiovese**.

Further evidence of the influence of Bordeaux is demonstrated by the choice of small, new **barrels** (*barriques*) for aging. In 1992, legislation introduced **IGT** into the denomination pyramid and these renegade wines finally found a worthy home. Two years later, in 1994, the *Bolgheri* DOC modified its rules to encompass Cabernet and Merlot. That same year, the *Bolgheri Sassicaia* DOC was formalized: a wine with its own denomination. These days, the Super Tuscan category spans a wide spectrum of wines. For many US drinkers, it is a natural stepping stone that leads from California or France into Italy. The term may have been coined by wine writer Burt Anderson, or by Robert Parker, or perhaps even Nicholas Belfrage MW.

Tannins

Builder's tea? Raw artichoke? We all know the dryness, bitterness and astringency of tannins. Their concentration defines a wine's pucker power. Tannins are plant-derived polyphenols and are in skins, stems, and seeds. A lengthy maceration can mean more tannins. Whether they provide a corset or a skeleton may say more about you than about the tannins themselves. High tannins make for better **bottle**-aging, though they do polymerize into longer (softer) chains over time. The *Sangiovese* grape, known as **Brunello** in Montalcino, is a relatively tannic grape. Part of the tannin component in *Brunello* also comes from the obligatory aging in wood, and sometimes from a producer choosing to ferment in wood.

Tartufo

Truffles are no trifling matter round here. As of 2016, the Montalcino civic area includes **San Giovanni d'Asso**, famous for its truffles: *Tartufo Bianco dei Creti Senesi, lo Scorzone* and the *Tartufo Marzuolo*. There are over thirty **hectares** of truffle sites (a testimony to the absence of pollution in this area), two events (November and March) and an amusing interactive museum located in a fourteenth-century castle. The museum is the first of its kind and includes an *odoroma*, described as 'an absolute joyride for the sense of smell'. Add to your sight-seeing list, please.

Tastevin

The most bling of tasting accessories, a *tastevin* is a shallow silver saucer that, from the 17th century onwards, was an indispensable tool of the trade for **winemakers** and tasters in France before spreading to the rest of Europe. Nowadays, it is infinitely more decorative than functional and can be worn around the neck on a chunky silver chain. The different indentations on the rim help to oxygenate wine quickly and catch maximum light. It can be used for either red or white wines depending on what hand you hold it in to tilt.

Silver does not affect the taste of the wine: an added plus. It was designed to be transportable and permit a rapid appreciation of color and clarity before better **glassware** and electric light made it redundant. It is the acknowledged ceremonial symbol of Burgundy and the official symbol of the **AIS**.

Teatro degli Astrusi
Montalcino's theatre overlooks *Piazza Padella*. It was built in 1766 by architect Leonardo De Vegni. At one point it was privately owned as a nightclub. After the successes of the *Festival Internazionale dell'Attore* it was purchased by the town council.

Temperini, Fedele
Fedele Temperini was born at *Poggio Alle Mura* (the original name for the castle now owned by Banfi) in 1892. He died in 1918, whilst shielding Ernest **Hemingway** (1899-1961) from mortar shells from the Austrian Line. At the time, Hemingway was a Red Cross volunteer distributing cigarettes and chocolates to the soldiers, and the name of the young man who had taken the brunt of the blast was unknown.

In 2019, thanks to joint research by US and Italian academics, the character Passini in *A Farewell to Arms* (1929) was identified as the young Montalcino resident and a plaque was hung near the WWI memorial under the *Cappellone* commemorating this brave twenty-six-year-old. After over one hundred years of anonymity, he deserves his own entry, as opposed to being a perennial footnote in history.

Tempio del Brunello
This interactive, immersive space is part of the museum complex run by

Ora di Montalcino and is a great starting place for a first-time visit to Montalcino. Don't miss the subtitled documentary.

Terroir
An overused word, and a challenge both to pronounce and to define. This French noun can be loosely interpreted as the interaction between topographical and geological features (such as soil, elevation and rivers), and climate, hours of sun or rain, and human intervention; it is what gives wines their sense of place. It used to belong to Old World wines, but now is applied to wines from anywhere (and Old World is also on the way out, given the colonial assumptions upon which it is based).

Tonneaux
Another barrel dimension, varying from 300 to 750 liters. So, basically: a big small barrel or a small big barrel.

Traditional
A word we love in Montalcino. Traditional generally refers to those producers who use large **Slavonian oak** *botti* (as opposed to small *barriques*) to age *Brunello*, and may veer towards keeping the wine in wood for the original time of four years, rather than the minimum two years. It is fair to say that producers who identify themselves as traditional are obsessive about fruit health and vineyard management.

Tramontana
This cool **wind** that comes in from the Alps can be vital to and desirable for drying out the grapes after rain and before harvest. In Montalcino, the *Tramontana* often heralds good weather. Great for laundry.

Travaglio
This Montalcino *quartiere* is identified by a yellow and red flag. On the last Sunday of May, *Travaglio* celebrates San Donnolo Donnoli.

Travaso
From the verb *travasare*, this literally translates as to transfer from one vessel to another so can refer to any movement of wine from barrel to barrel. The verb is also used for re-potting plants.

Trescone
This ancient Tuscan group dance is performed to an accordion and involves lots of floppy-hat-tossing, foot-stamping and partner-swapping. The folk dance is mentioned by both Dante and Boccaccio (who visited Montalcino in 1136) and features in the original *Pinocchio*. The *Trescone* has been part of life in Montalcino since 1941. Montalcino's youngsters are still recruited to dance at various moments throughout the year: **Sagra del Tordo**, *Sagra del Galletto*, **Apertura delle Cacce** and **Santo Patrono.**

UFO

In the *Pianello* quarter stands the Romanesque church of San Lorenzo in San Pietro, once a dependence of the abbey of *Sant'Antimo*. Dating back to the fourteenth century, it contains three paintings by Bonaventura Salimbeni (1567-1613), one of which, the *Glorification of the Eucharist*, was thought to depict a UFO. This came under intense scrutiny from the UFO community in the 1980s and was described as the 'Sputnik of Montalcino'. It does, indeed, show a strangely modern sphere with wand-like antennae, but it is now thought that this object is just a globe.

Ullage

This is the gap between the cork and the wine in the **bottle**. As time passes, some wine is inevitably lost through evaporation and cork absorption. Ullage is more evident in older wines and can be an indication of storage conditions. Also known as fill level or headspace. Rhymes with Dulwich.

Ungulates

This is the family of even-toed herbivorous mammals that includes both deer and boar, both of which pose huge threats to Montalcino vineyards and have resulted in ubiquitous **fencing**. The word 'ungulate' comes from the Latin 'ungula', meaning hoof. Rather bizarrely, recent research indicates that dolphins and whales also belong to this family though – so far – they pose no risk to vineyards.

Unification of Italy

This process was the result of nearly sixty years of events and revolutionary ideas that culminated in 1861 when the Italian state was born and in 1871 when Rome was declared capital of the Kingdom of Italy. Until then, Italy had been a complicated jigsaw puzzle of kingdoms and city-states. In Montalcino, as in other parts of Italy, the unification led to the renaming of many squares and streets.

Mazzini, Italian patron and politician, together with Giuseppe Garibaldi, Camillo Benso di Cavour and King Victor Emmanuel II, was one of the leading players in the Italian *Risorgimento*. The main street of Montalcino is called *Via Mazzini* (you will find one of these in almost all Italian towns) after the hero of the *Risorgimento*. We lived at number 37 for five years. For the record, Mazzini (1805-1872) visited only three places in Tuscany: Florence, Livorno and Montalcino.

Val d'Orcia
The ***Val d'Orcia*** is one of Tuscany's seven UNESCO World Heritage sites and joined this prestigious list in 2004. The boundary of the World Heritage property coincides with the boundaries of the modern-day Park of *Val d'Orcia* (*Parco Artistico Naturale e Culturale della Val d'Orcia*). The UNESCO criteria are met by this area encapsulating the essence of Renaissance landscape, as exemplified by the *Scuola Senese* and Lorenzetti's painting in the town hall of Siena. It was the economic decline of the *Val d'Orcia* and the resulting poverty and marginalization that preserved the Renaissance layout and landscape aesthetic.

Val d'Orcia Gravel
This non-competitive international cycling event was introduced in 2022 as part of Montalcino's post-pandemic resurrection. It takes place in the third week of October, as the vines turn golden: a beautiful time to be on two wheels. There are four different routes, organized by a local association that has been dedicated to promoting '*cicloturismo*' since 1990. The shortest route can be enjoyed on an e-bike and includes lunch at a winery.

Valoritalia
This national body is authorized by the MiPAAF (Ministry of Agricultural, Food and Forestry Policies), rebaptized in 2023 as the grandiose MASAF (Ministry of Alimentary Sovereignty and Forests – thanks, Giorgia Meloni) to certify and watch over Italy's **denomination** wines. This third-party entity became the controlling body for **Brunello** in 2009 after **Brunellogate**, and guarantees traceability at every stage of production. In 1979 the ***Consorzio del Brunello*** had been given the powers of *vigilanza* and *tutela*.

Vandalism

Montalcino, sadly, is no stranger to vandalism. If only it were limited to the playful repositioning of the figures in the ***presepe*** (nativity scene), night after night before Christmas. There are well-known tales of tractor tires burned in proximity to a neighbor's vineyard just before harvest, resulting in rubbery smoky tastes in the subsequent vintage. In some areas of Montalcino, multiple producers are side by side and not everyone can avail themselves of the same level of technology. Batteries used to power electric fences that dissuade wild boar from invading vineyards might be accidentally switched off by neighbors without electric fences in order to level the playing fields. Gates and fencing can be sabotaged. One year, a whole vineyard of ***barbatelle*** (baby-vines) was hewn overnight with a chainsaw in the very year in which they would have started producing fruit that could have been harvested, three years into the process.

Most infamous of all was the willful emptying of six vintages of production (circa 62,600 liters – vintages 2007-2012 – ten **barrels** down the drain) of iconic producer Soldera in 2012.

Vegan

Food for thought: organic and biodynamic vineyards have high insect **MOG** and often rely on animal-derived fertilizers. Cow Horn 500, I am looking at you. Does this mean that truly vegan wine risks ending up being heavily reliant on pesticides and chemicals? Probably not, but, as always, things are more complicated than they seem.

Vendemmia

This refers to harvest as in the actual picking of the grapes. The word for vintage is *annata*. In Montalcino, owing to the difference in **altitude** between low-lying and higher vineyards, the generalised picking period may take as long as a month, from early September onwards. This happened, memorably, in 2017. If you are visiting in autumn, be sure to calculate extra travel time, since many wineries are transporting, via tractor, the crates of grapes from vineyards located in different parts of the hill back to the centralised cellar. In the period before harvest, Montalcino is animated by a shared frenzy and it is the main topic of all conversations, at the school gate, at the beauticians or in the bars.

Veraison
In Italian: ***invaiatura***. This is the moment when grapes begin to change color, size and consistency, and is an important phase in ripening.

Vermiglio
Vermiglio is an ancient Italian word, used by both Boccaccio and Dante Alighieri, to describe a type of vibrant or bright **red**. There was once a wine called *Vermiglio* in Montalcino with a short cellar life and wonderful freshness and fragrance. Could this be claimed as a precursor of **Rosso di Montalcino**?

Via Francigena
This is the 3,200 km pilgrim route linking Canterbury to Rome, that created and enforced communication and trade links in medieval times. It goes past Torrenieri (Tappa 35) and millions of pilgrims and travelers heading to Rome inevitably passed by Montalcino. In the thirteenth century, the town acquired the right of *Porto Franco*, becoming a sort of duty-free area since Montalcino could determine its own taxation. This religious tourism, which included kings, popes and cardinals, caused a surge in the production of Montalcino wine, which at the time consisted of **Moscadello**, a sweet white wine.

If you have spotted people laden with rucksacks plodding along beside the road, it is most likely that they are doing some or all of the *Via Francigena*. In 2022, fifty thousand people, hailing from forty different countries, walked this route. In recent years, footpaths and bridges have appeared to accommodate these modern pilgrims.

Vigna

The **Brunello disciplinare** has an even more exacting set of rules that govern single vineyard production. If the word *Vigna* or *Vigneto* appears on a *Brunello* label, this means that the wine must have had a traceable separate path from fermentation onwards. The **yield** is lower and the **ABV** should be higher. **Rosso di Montalcino** can also exist in *Vigna* expression, reflected by a similar series of tightened legislation regarding yield, ABV and traceability.

The first single vineyard wines in Montalcino date back to the mid 1970s and the Montosoli area. These days, at least 15% of *Brunello* is bottled with a site-specific name. Tasting these expressions can be enlightening: a shortcut to understanding the intricacies of the many Montalcino micro-*terroirs*.

Vignaiolo

This word, in Italian, comes trailing with significance. Just like the French term '*vigneron*', the word for vine (*vigna/vigne*) is at the root of it all. The best translation is 'grower'. *Vignaioli* are the people who grow grapes and this underlines the significance that vineyard management has in making high quality wine. The opposite, sometimes derogatory term, would be vintner or **commerciante** – a merchant, someone who trades in wine but is divorced from the cultivation of the vine.

Vimini

Willow twigs, known as *vimini* or *vinco*, were once used for tying up the vines, after being soaked for flexibility. Some wineries are experimenting with reviving this technique as an alternative to plastic ties.

Vinaccia

Marcs, or pomace, are a byproduct of winemaking. After fermentation and pressing, the skins and seeds form a cake. These moist, fragrant marcs can be distilled to make **grappa**. The *-accio* **suffix** is pejorative and indicates something nasty. *Vinaccia* is a color descriptor in Italian to indicate what in

English we refer to as burgundy (which in Italian is *bordò* from Bordeaux). Take a moment to think this through.

Vini da Meditazione

No yoga mats here. This term was coined in the 1970s by **winemaker** Luigi Veronelli (1926-2004) to do justice to those wines that give their best when drunk unaccompanied by food: either sweet full-bodied wines with intense and complex aromas suitable to end a meal or great red wines with age to them. Their ideal pairing is with time and patience, in order to appreciate fully all their nuances and delights.

Vintage Evaluation

It ain't over 'til it's over. Producers are generally very reluctant make pronouncements about the vintage during the course of the year. It is partly in order not to invite bad luck, but it is also true that vintages often begin one way and finish another and it is perfectly possible to lose a year of work and a year of future income in a five-minute hailstorm. A vintage may have seemed more than promising before hopes are dashed, just as the year may appear initially uninspiring or even ill-fated and yet turn around in September. Tasting **must** or embryonic wines in **barrels** requires experience and a certain amount of mental gymnastics. It is rather like holding an acorn and envisaging the oak tree it will become. Of course, *Sangiovese* and *Brunello* experts can predict exactly that, and the wines' analytic parameters offer runic insights.

Until 2021 vintage, every vintage was given a star rating in the February following harvest. In fact, the attribution of the stars and the cementing of the celebrative tile, or *mattonella* was an integral part of **Benvenuto Brunello.** In order to decide on the rating, samples of wine pre- and post-fermentation were collected from different wineries in different areas of Montalcino and a series of experts was gathered to make the decision.

This system was abandoned from 2021 vintage onwards, due to the timing of *Benvenuto Brunello*, which in that year was held in November for the first time. It is unlikely that 2021 will ever be rated, nor the years thereafter, since the system is finally under review. The idea of rating vintages was always facile, scarcely objective and inevitably steered by a growing awareness of market repercussions. In addition, anyone familiar with Montalcino will know that vineyard location can change the experience of a growing year, due to **microclimate**, soils, difference in **altitude, exposition** or sheer good fortune. Awarding stars to a vintage negates and flattens the effects of the human decisions that wineries can make in challenging years, which can mean wildly different outcomes for neighboring producers. That said, it can be a good indication of **drinking window** and no one can deny the pleasure of realizing that one's own birth year was a great vintage.

Vintage of the Century
We have these quite often – 2010 was a monumental vintage in Montalcino, thanks to near perfect weather conditions in all compass points and **altitudes**. Generous winter and spring rains gave the vines vigor, followed by a warm summer with no damaging **heat spikes** (2011, I'm talking to you) and a great September distinguished by extreme day/night thermal excursion. Depth, complexity, structure and great **acidity**: the perfect recipe for longevity and delightful drinking. For five years, 2010 was hailed as the vintage of the century, before losing the title to 2015 and then 2019.

VIPs
Montalcino is much beloved as a destination for the famous and wealthy. Michelle Obama had ice cream here in 2017, and the Biebers are regular visitors to the area. When the US basketball player, LeBron James, mentions a *Brunello* on his Instagram feed, the internet breaks. If you think you might be sitting next to a Hollywood star while you're having a morning coffee here, you probably are.

Films, adverts and series are filmed in the environs, while the stations at *Monte Amiata Scalo* and *Sant'Angelo Scalo* are often transformed into period sets.

Volatile

In Italian (be sure to pronounce as four syllables), this is VA (Volatile **Acidity**), the measure of gaseous acidity in a wine and an indicator of spoilage, poor quality grapes and/or unhygienic cellar conditions. To be clear, tartaric and malic acids are non-volatile but acetic acid is volatile.

Wars
Every village has a list of their war dead, affixed in the main squares. In Montalcino, the Great War dead are commemorated on one of the walls of the **Cappellone** and there is a memorial in the **Giardini del Impero** to those lost in World War Two, along with various memorial stones like the one in **Nacciarello** where citizens lost their lives. For a gripping account of the Second World War in this area, I recommend *War in Val d'Orcia: An Italian War Diary 1943-1944* by Iris Origo or, for Italian speakers, Ivo Caprioli's *La Liberazione di Montalcino*. My father-in-law was exiled from Sant'Angelo in Colle to a remote farmhouse. He told of different meal sittings for the women and children, the latter's faces often thick with flies since the farmhouse kept livestock on the bottom floor for warmth.

Whale in a Vineyard
The unimaginatively named 'Brunella' is one of the few fossils to have her own hashtag. This prehistoric whale, between six and eight meters long and with an estimated weight of around six tons, is a tangible and indisputable example of the presence of marine fossils and all that this means for *terroir* and grapes. More than a decade after her discovery, this wonderful four-million-years-old creature has been restored and reassembled by an expert team. The discovery of shark teeth embedded in her vertebra suggest that she came to a violent end. She is considered to be one of the best-preserved Mediterranean fossil specimens.

Wind
Some names of farmhouses in this area – for example, *Sparaventorio* or *Ventolaio* or the winery *Corte dei Venti* – give a clue to the omnipresence

of wind in Montalcino, hardly surprising given its position and **altitude**. If you have ever turned a street corner in November only to find yourself blasted off your feet and chilled to the bone, you will immediately recognize the role of *il vento* in Montalcino. Many wineries claim their **exposition** to wind is an essential aspect of their *terroir*, given the role of the wind in slowing of photosynthesis, drying out of humidity during maturation or pre-harvest, or even increasing *stress idrico* in summer. The main winds that blow through and around Montalcino are *Tramontana, Libeccio, Scirocco, Grecale* and *Maestrale*. These names come from the historical navigation charts of the Venetians. These able mariners used the island of Malta as a reference point and the names of the Mediterranean winds are based on their direction with respect to Malta – just in case the compass points seem a little off. The most fearsome of all winds is the *Burian* or *Buran*, an icy wind from Russia that brings incredibly low temperatures. It struck Montalcino in February 2021.

Winemaker
This role of the *enologo* is less rockstar than in the US, since many wineries in Montalcino 'share' the same consulting winemakers rather than having a dedicated member of staff. If anything, due to climate change, the roles of the *agronomo* and *fattore* are growing in importance.

Winenews
This communication agency was founded by an enterprising Montalcino couple, Alessandro Regoli and Irene Chiari. In May 2000, winenews.it went online and has been supplying daily updates to the wine world on a multitude of platforms ever since. It is a great database and source of information about the world of wine and food.

Winkler Index/Scale
I know, right. This is a 1940s US system for grading regions according to average temperatures. It was created after Prohibition to help producers

decide which grapes to plant where. It's superannuated and insufficiently specific and yet always provides the opportunity for a knowing smirk.

Wolf Statue

The white marble column was erected in April 2012 between **Piazza Padella** and **Teatro degli Astrusi**. Sometimes it is adorned with flags that celebrate the original three town divisions (*terzieri*) rather than the actual town quarters (*quartieri*). It is a replica of the original column that was in that exact position until 1618 (when it was toppled and the wolf was decapitated). You can admire the headless wolf statue in the *Museo di Arte Sacra*. Should you be confused, the wolf and the suckling babes are a symbol of Siena as well as of Rome. Legend has it that Siena and Asciano were founded by Remus' two sons, Aschio and Senio, who escaped from their power-mad fratricidal uncle Romulus in Rome. In seventeenth-century Montalcino, the column was a reminder of unwanted Sienese dominion and was presumably desecrated by Florence supporters. Although the replacement is still standing and whole, it is not a beloved monument. Hardly anyone turned up to the inauguration and there was public grumbling about the expense.

Wolf Urine

This is one way to banish wild boar or *cinghiali*. Small and surprisingly expensive bottles of repellent can be sourced from Spain. Apparently, wolf urine works for raccoon too, according to US comedian Aaron Weber. Other methods include rags soaked in gasoline, human hair and Marseille soap in socks hung from the vine posts. Or a fence.

Wolves

Yes, we have them. First sightings, coincidentally, date from 2012, the year the **Wolf Statue** was erected. Their presence is very problematic for the shepherds who make pecorino cheese.

Xylella fastidiosa

Scrabble players, this is my gift to you. *Xylella fastidiosa* is a terrible bacterium, transmitted by insect vectors. It can attack six hundred plant species belonging to more than eighty different botanical families.

In Italy, in 2013, it devasted olive trees, oleanders and almond trees in Puglia. There is no cure for infection and European regulations advocate uprooting and destroying contaminated plants.

Yeasts

Yeasts are an unexpectedly sexy subject in Montalcino. The choice is between selected yeasts and indigenous yeasts and it is worth knowing that you should react with jaw-dropping awe when the latter are mentioned. The words 'native' and 'wild' are used interchangeably and with abandon. Indigenous or ambient yeasts are the yeasts that are naturally occurring in the vineyard, so represent the apex of local, *terroir*-specific winemaking: the sourdough of grape fermentation. Selected yeasts are their tedious, predictable cousins, that can be bought in a bag and offer reliability but less frisson of excitement and potential for anarchy.

Yields

The maximum yield, or *resa*, for **Brunello**, per **hectare** and per vine, is specified in the **DOCG** regulations for this wine. Yield can be affected by weather: a late freeze as happened in 2017 or 2021, humidity during flowering, or summer heat and **drought** – or even all of the above. Yield can also be conditioned by the producer's own canopy decisions and quality goals. Simply put, all being well, less is more, ie the fewer bunches on the vine, the more share of nutrients they will receive and therefore be improved as a result. The **density** of the vineyard, ie how many plants per row, and the *potatura* methods will also determine potential yield. These days, a 'normal' vineyard will be 2.5 meters between rows and 80cm between the plants, making for a density of 5,000 vines per hectare.

Zafferano

Saffron has been cultivated in the **Val d'Orcia** since the Middle Ages and has recently been revived in Montalcino. In 1857, at the *Esposizione Agraria Toscana*, Clemente Santi (yes, that Santi) presented the *Zafferano del Suolo Montalcinese* some twelve years before he presented a wine he called *Rosso Scelto (**Brunello**)* vintage 1865.

Zappatura

A *zappa* is a hand-held hoe; *zappatura* is the act of hoeing the soil directly around the growing vine. This manual operation cannot be substituted by machine operations such as *scavallatura*. There is a lovely Italian expression for inadvertent self-harm: *darsi la zappa sui piedi* – the equivalent of shooting yourself in the foot or cutting the branch on which you are sitting – rather like the act of writing an idiosyncratic and frivolous primer about the town in which you have lived and worked for almost thirty years.

Vintage Evaluation

Until vintage 2021, every vintage of **Brunello** was awarded a rating out of five stars while the wines in question were still in barrel. From 1992, an influential personage was commissioned to design a ceramic tile (*mattonella*) to commemorate the vintage. These tiles are officially presented with a deal of fanfare and in the presence of the creator, the **Consorzio del Brunello** and all the town grandees. The tile is cemented into its place on a wall to be a forever memento of the vintage.

Of course, the best place to see these tiles is *in loco*. Come to Montalcino. Find **Piazza Padella**. Make your way to the marble bench in front of the **Wolf Statue**. Perhaps you will be squeezed between two local ladies having a rest in the shade, perhaps you will find a moment when you are there alone. Take your time to look at the wall of the **Palazzo dei Priori** where the tiles are embedded. Failing that, they are easy to consult on the *Consorzio del Brunello* website, under Vintage Quality Evaluation. The *mattonelle* range from the beautiful to the self-promotional. They are thoughtful – or, to be honest, sometimes, a little bit lazy. The subjects and the choice of author often offer an insight to the *zeitgeist* of the year in which they were commissioned.

Below is the list of celebrity tile-designers, along with the vintage ratings, by year.

2023 Monica Maggioni, Italian journalist and war reporter, executed by Dario Curatolo. The subject is the collective drama of war that is playing out beyond Montalcino.

2022 Fashion designer Brunello Cucinelli, who now also produces wine in Umbria. It says '*Al Brunello di Montalcino il celebre vino, ove mi risconosco per nome avvicina alla Sapienza di Dionisio*' – making a joke of his sharing his first name with the wine and it being close to the wisdom of **Dionysus**.

2021 Chef Carlo Cracco. It shows an egg: symbol of life and associated with Cracco, famous for his crispy egg yolk amongst other things. The background is Galleria Vittorio Emanuele in Milan, site of his Michelin-starred restaurant. This choice was in recognition of Cracco's personal contribution to elevating all things Italian and of how hard the restaurant sector was affected during the pandemic, with a 40% decrease in turnover.

2020 ★ ★ ★ ★ ★
Olympic champion swimmer Federica Pellegrini, showing a phoenix, just like her neck tattoo. The choice was particularly appropriate given the pandemic that was ongoing at the time. It reads *'ad ogni vendemmia la rinascita di un mito'* – 'every harvest a myth is reborn'.

2019 ★ ★ ★ ★ ★
Giovanni Malagò, President of the Italian National Olympic Committee.

2018 ★ ★ ★ ★
Paralympic athlete Alex Zanardi. Take the time to make out the phrases written in the stars.

2017 ★ ★ ★ ★
Sting; similar to one of the labels he produced in his winery in Valdarno.

2016 ★ ★ ★ ★ ★
Commemorating the relationship with Michelin Guide; ceased in 2023.

2015 ★ ★ ★ ★
Featuring five artists, celebrating five decades since **Brunello** became one of the first **DOC** wines in Italy. The artists are ceramics duo Bertozzi & Casoni, Sandro Chia (see **Mosaic Roundabout** and the 1994 tile), Pino Deodato, Gian Marco Montesano and Mimmo Paladino.

2014 ★ ★ ★
Carlo Petrini, activist; founder of the International Slow Food movement.

2013 ★ ★ ★ ★
Oscar Farinetti, founder and owner at the time of Eataly and, less relevant to Montalcino, the electronics chain Unieuro.

2012 ★ ★ ★ ★ ★
Cruciani, the name behind the Made in Italy *macramè* bracelets that were all the rage in 2012: even worn by Psy in the *Gangnam Style* video.

2011 ★ ★ ★ ★
Salvatore Ferragamo. The Ferragamo family have close historical links with Montalcino. In 2022 they sold their winery, Castiglion del Bosco.

2010 ★ ★ ★ ★ ★
Celebrating 150 years of the **Unification of Italy** and *Brunello* (1861-2011).

2009 ★ ★ ★ ★
Tadashi Agi and Okimoto Shu, authors of the popular manga, *Drops of God*, made into a series in 2009 and then again in 2023.

2008 ★ ★ ★ ★
In 2007, Alessandro Grazi, an artist from Siena, was commissioned to design the *Palio drappellone* (a painted silk cloth given to the *contrada* that won the horse race). It was no surprise when the **Consorzio** enlisted him to design the 2008 tile.

2007 ★ ★ ★ ★ ★
Roberto Giolito, car designer at FIAT. The tile features the beloved Fiat 500 that was relaunched in 2007.

2006 ★ ★ ★ ★ ★
Adam Tihany: hotel/restaurant designer and architect, generally credited with creating the concept of the upscale trattoria and brasserie in New York. Born in Transylvania in 1948, he grew up in Israel and studied architecture in Italy before moving to the US.

2005 ★ ★ ★ ★
Riccardo Benvenuti, a Tuscan painter, known for his portraits of women.

2004 ★ ★ ★ ★ ★
Peter Weller, yes, the *Robocop* actor.

2003 ★ ★ ★ ★
In homage to Sienese painter, Duccio di Buoninsegna (c.1255-1318), the tile shows the *Nozze di Cana predella* where water is turned into wine. In 2003, there was a wonderful Duccio exhibition in Siena.

2002 ★ ★
Roberto Cavalli, Italian fashion designer.

2001 ★ ★ ★ ★
Miuccia Prada, fashion designer and businesswoman. The leaves shown are most definitely not **Sangiovese**.

2000 ★ ★ ★
Emilio Giannelli, a famous Italian cartoonist from Siena, combines the tongue action that all Italians associate with the film character *Fantozzi* with *La Gioconda*, who is clutching a **bottle** of **Brunello**.

1999 ★ ★ ★ ★
Giorgetto Giugiaro, nominated Car Designer of the Century in 1999.

1998 ★ ★ ★ ★

Immediately identifiable as work by Ottavio Missoni (1921-2013) from the homonymous fashion house.

1997 ★ ★ ★ ★ ★

Deborah Compagnoni, Italian ski champion with three gold medals at the Olympics and as many in the World Championships.

1996 ★ ★ ★

Pierluigi Olla, a Sienese sculptor who designed the latest *Sagra del Tordo* costumes. In Siena, look out for his work in Via dei Rossi, showing a woman at a window.

1995 ★ ★ ★ ★ ★

At the time this was immediately recognizable as being by photographer Oliviero Toscani, famous for his controversial ad campaign for Benetton between 1982 and 2000.

1994 ★ ★ ★ ★

Sandro Chia, artist painter and sculptor and renowned member of the *Transavanguardia* movement. He also contributed to the 2015 tile and is the artist behind the **Mosaic Roundabout.** He owns the Castello Romitorio winery in Montalcino.

1993 ★ ★ ★ ★

Paul Leber was a Swiss painter (1928-2015).

1992 ★ ★

Local artist Roberto Turchi created this first tile. His daughter has a wonderful fashion brand, Clotilde, well worth checking out.

1991 ★ ★ ★
1990 ★ ★ ★ ★ ★
1989 ★ ★
1988 ★ ★ ★ ★ ★
1987 ★ ★ ★
1986 ★ ★ ★
1985 ★ ★ ★ ★
1984 ★
1983 ★ ★ ★ ★
1982 ★ ★ ★ ★
1981 ★ ★ ★
1980 ★ ★ ★ ★
1979 ★ ★ ★ ★
1978 ★ ★ ★ ★
1977 ★ ★ ★ ★
1976 ★
1975 ★ ★ ★ ★ ★
1974 ★ ★
1973 ★ ★ ★
1972 ★
1971 ★ ★ ★
1970 ★ ★ ★ ★ ★
1969 ★ ★
1968 ★ ★ ★
1967 ★ ★ ★ ★
1966 ★ ★ ★ ★
1965 ★ ★ ★ ★
1964 ★ ★ ★ ★ ★
1963 ★ ★ ★
1962 ★ ★ ★ ★
1961 ★ ★ ★ ★ ★
1960 ★ ★ ★

1959 ★ ★ ★
1958 ★ ★ ★ ★
1957 ★ ★ ★ ★
1956 ★ ★
1955 ★ ★ ★ ★ ★
1954 ★ ★
1953 ★ ★ ★
1952 ★ ★ ★
1951 ★ ★ ★ ★
1950 ★ ★ ★ ★
1949 ★ ★ ★
1948 ★ ★
1947 ★ ★ ★ ★
1946 ★ ★ ★ ★
1945 ★ ★ ★ ★ ★

Some Dates in the Montalcino Calendar

Check the noticeboards and the Proloco and the Oro di Montalcino website. Fridays are market days; consider yourself warned: parking is a challenge.

February	***Brunello Crossing***
March	First/second Saturday: ***Strade Bianche*** Fourth weekend: *Festa del Tartufo Marzuolo* in **San Giovanni d'Asso**
May	8th: ***Santo Patrono*** ***Eroica*** Last Sunday: ***Travaglio*** celebrates San Donnolo Donnoli ***Pianello*** honors San Pietro
June	**Red Montalcino**
July	Jazz & Wine in the ***Fortezza*** Last Sunday: ***Ruga*** celebrates San Salvatore
August	Second weekend: ***Apertura delle Cacce*** Keep an eye out for **Camigliano** Blues dates
September	***Settimana del Miele*** First Sunday: ***Borghetto*** commemorates Sant'Egidio
October	First Sunday: *Sagra del Galletto* in **Camigliano** ***Val d'Orcia Gravel*** Last weekend of the month, ***Sagra del Tordo*** in Montalcino
November	***Benvenuto Brunello***

Further Reading in English

Christy Campbell: *Phylloxera – How Wine Was Saved for the World* (2004)

Stefano Cinelli Colombini; translated by Jeremy Parzen:
Brunello di Montalcino: portraits from memory
The history, places, and people behind the legend (2020)

Isabella Dusi: *Vanilla Beans and Brodo* (2001)

Ferenc Máté: *A Vineyard in Tuscany* (2007)

Alexander McCall Smith: *My Italian Bulldozer* (2016)

Kerin O'Keefe: *Brunello di Montalcino: Understanding and Appreciating One of Italy's Greatest Wines* (2012)

Kerin O'Keefe: *Franco Biondi Santi: The Gentleman of Brunello* (2005)

Iris Origo: *War in Val d'Orcia: An Italian War Diary 1943-1944* (1947)

Afterword

Grand Duke of Tuscany Peter Leopold of Lorraine (1747-1792), brother to Marie Antoinette, observed in his travels, 'one of the great passions of Montalcino is to have all the things in a small way like in Siena.' My experience of this Tuscan town is that it is quite unlike anywhere else. This is a feeling shared by nearly all of those who are born there, and by many who join the community. In fairness, Siena feels exactly the same way, but that is another matter. 'Wha's Like Us?' *campanilismo* is strong in this area.

In the *Museo del Brunello*, there is a whole room entitled *Stanza dei Primati*, the Room of First Records and not a monkey in sight. Here you will find information regarding the oldest wine label to be printed in Italy, the first cellar visits, the first wine to be sold by correspondence and more. Mayor *Ilio Raffaelli* was also very fond of thinking of Montalcino in terms of its firsts, independently of whether they took place during his own tenure. He afforded the same pride in talking about Montalcino's early adoption of electric lights in 1902 via a steam plant powered by charcoal as to it having the first full-time school in Italy almost seventy years later under his aegis.

Montalcino was my first love and my home for many years. I exited our flat on *Via Mazzini* in a beetle-green wedding dress in 2000 to marry my Super Tuscan and pushed prams up and down its streets in the years thereafter. In truth, this love affair began before I was even born. My newlywed parents accepted an invitation to spend some of their summer at a far flung *podere* way below Montalcino, provided they brought their own airbed, plate and cutlery. The young Mr and Mrs Gray were exposing me to Italophilia well before my arrival in this world. They bought a small home in Camigliano and were amongst the first *stranieri* in this area. Apparently, I was so accustomed to the treatment afforded babies in Italy by virtue of simply being babies, that back in my pram in bleak Scotland they could see me looking around expectantly for a vituperative *nonna* to pinch my cheek.

As I grew older, the time I spent in Italy became increasingly special, not least to do with a subtle shift in parental rules and regulations as soon as we arrived. Very quickly, a divide was established; school in Scotland, walking there and back in the dark, grey months punctuated with exams and music lessons. In contrast, our long family summers in Tuscany were aglow with fireflies and shooting stars, river swims and thick pear juice on market days in the times when the *Fiaschetteria Italiana* still had stuffed songbirds on display in its windows. After years of this schizophrenic existence, it was only natural that, after finishing my studies in Britain, I would jump off the golden path to hurl myself towards my real purpose in Italy. Only now that I am a parent myself can I understand how difficult it must have been to witness this apparently kamikaze behavior.

I was working in a *pizzeria* in Paganico and driving a pistachio-colored Renault 4. I was translating, giving English lessons and slicing bread and *prosciutto* for bikers in the small kitchen in Marco's bar in Sant'Angelo in Colle. Word got out that there was a person who spoke two languages and could touch type. Bear in mind that this was 1995 and these two attributes made me a rare creature. I received a sheaf of offers and ended up working for Paola Gloder, then owner of the *Poggio Antico* winery. Like anyone who is good at everything, she was an exacting boss. I remember shedding tears as she hammered double entry bookkeeping (the Medicis' most lasting legacy) into my English-graduate brain. But learn I did, and after my seven-year stint with her, I was also a dab hand at loading trucks and manipulating Excel formulas. This stood me in good stead for my time working for two wineries, *Terralsole* and *Il Palazzone*. In 2007 I began to work solely for *Il Palazzone*. These were the blurry years of having a young family. We lived at *Le Due Porte* for eight years, during which time our third child was born. In 2015 we moved to Siena, thrust by internal and external forces. School beckoned, along with a wider world and the excitement of being able to fill a shopping trolley without observation or comment.

Every time I drive up through the fog to Montalcino, I am overcome by its beauty. It makes me very happy that I continue to have many reasons to make that trip.

Thank you for having me, *Mons Ilex*!

Acknowledgements

You are probably familiar with the phrase 'It takes a village'. It did.

A few entries appeared in earlier versions on the *Il Palazzone* blog or Instagram feed during my time there. Many thanks to Peter and Kirsten Kern for consenting to their reproduction here.

The thankless and incredibly tedious task of reading this volume in one sitting fell to my early readers. Many thanks to Esther Juergens and Tim Heaton for their kindness and precious input, and to Emily O'Hare for taking the time to engage with this project.

I would like to recognize the untiring commitment of Francesco Belviso, Montalcino's unofficial photographer and town chronicler, who watermarks his photographs with his nickname, Biba. We are all similarly indebted to Monty Waldin who keeps an updated dictionary of wine on his blog, chateaumonty.com.

I am very grateful to Jeremy Parzen, aka Do Bianchi, for kind permission to share his *Sovescio* research. It is a pleasure to signpost his excellent online Wine Glossary which can be accessed via his blog. The generosity of Stefano Cinelli Colombini in receiving me and sharing his vast and far-reaching knowledge about Montalcino was remarkable. Similarly, I am indebted to Gabriele Gorelli, Italy's first Master of Wine, for agreeing to write the introduction, to Paola Martino for her oenological input and to Giovanni Stella for his agronomical expertise and constant kindness.

Many thanks also to Katja Meier and Alexandra Korey, who both provided vital assistance in crucial moments. I am much obliged to Montalcino-born graphic designer Isotta Rabissi, who created my logo, and to Kirstie McConnell, for her generosity and photographic skills.

Mille grazie to my wonderful editors, Jill Glenn and Claire Steele, for helping me get this out of my head and onto paper and for 'seeing' it before I did myself. And, of course, a big thank you to my parents for taking me to Montalcino in the first place, and to my husband for giving me a reason to stay.

Constellations Press is a small independent press committed to publishing works of fiction, memoir and essays.

We publish books that boldly reimagine society and celebrate our diverse humanity, adding to the total sum of the world's beauty.

constellationspress.co.uk

www.ingramcontent.com/pod-product-compliance
Lightning Source LLC
Chambersburg PA
CBHW030108240426
43661CB00031B/1341/J